THE IDEALIST

AARON SWARTZ AND

THE RISE OF

FREE CULTURE

ON THE INTERNET

JUSTIN PETERS

SCRIBNER   New York   London   Toronto   Sydney   New Delhi

SCRIBNER
An Imprint of Simon & Schuster, Inc.
1230 Avenue of the Americas
New York, NY 10020

First Scribner hardcover edition January 2016

SCRIBNER and design are registered trademarks of
The Gale Group, Inc., used under license by Simon & Schuster, Inc.,
the publisher of this work.

For information about special discounts for bulk purchases,
please contact Simon & Schuster Special Sales at 1-866-506-1949
or business@simonandschuster.com.

The Simon & Schuster Speakers Bureau can bring authors to your
live event. For more information or to book an event, contact the
Simon & Schuster Speakers Bureau at 1-866-248-3049 or visit our
website at www.simonspeakers.com.

Interior design by Kyle Kabel

Manufactured in the United States of America

10   9   8   7   6   5   4   3   2   1

ISBN 978-1-4767-6772-7
ISBN 978-1-4767-6773-4 (ebook)

To my parents

CONTENTS

Author's Note   *ix*

INTRODUCTION   The Bad Thing   *1*

CHAPTER 1   Noah Webster and the Movement for Copyright in America   *17*

CHAPTER 2   A Tax on Knowledge   *41*

CHAPTER 3   A Copyright of the Future, a Library of the Future   *67*

CHAPTER 4   The Infinite Librarian   *93*

CHAPTER 5   The Case for the Public Domain   *117*

CHAPTER 6   "Co-opt or Destroy"   *143*

CHAPTER 7   Guerilla Open Access   *169*

CHAPTER 8   Hacks and Hackers   *195*

CHAPTER 9   The Web Is Yours   *221*

CHAPTER 10   How to Save the World   *247*

Acknowledgments   *271*
Notes   *273*
Select Bibliography   *309*
Index   *319*

AUTHOR'S NOTE

This book began as an article in the online magazine *Slate*. The article was a long profile of Aaron Swartz, his life and work and undoing, written and reported in the weeks after his death. Called "The Idealist," it ran on February 7, 2013. The next day, an editor at Scribner contacted me and suggested that I consider expanding the story into a book. I spent several months completing a book proposal, and when I was finally finished, I realized that my book had no title. After giving it some very brief consideration—I pluralized the title of the *Slate* profile and called the book *The Idealists*, thinking that we would come up with a better title later. We did: *The Idealist*. Sometimes your first instincts are your best ones.

The Idealist is not meant as a comprehensive biography of Aaron Swartz or a comprehensive history of Internet activism or American copyright law. Little at all about this book is comprehensive. Someone could easily make another book consisting exclusively of material omitted from this one, and if you do, please send me a copy. Instead, *The Idealist* is a provisional narrative introduction to the story of free culture in America, using Swartz's life as a lens on the rise of information sharing in the digital age.

The first half of this book presents a basic narrative introduction to the history of copyright and free culture in America. In these chapters, I focus on various representative individuals whom I found interesting and significant, in hopes that emphasizing their personal stories will help animate and simplify otherwise abstract topics. Chapters 3 and 4 utilize archival research performed at the University of Illinois at Urbana-Champaign, which hosts both Michael Hart's personal papers and the archives of the American Library Association.

The second half of this book is based mostly on a close reading of Swartz's own writings, supplemented by interviews and public documents. Chapter 8 employs a trove of redacted internal JSTOR e-mails to tell the story of Swartz's fateful MIT downloading spree. Chapters 9 and 10 draw on thousands of pages of redacted US Secret Service documents acquired through the Freedom of Information Act by *Wired*'s Kevin Poulsen. Throughout the book, I have occasionally reused some material from my original *Slate* profile of Swartz. This material is used with *Slate*'s permission, and I am grateful to *Slate* for granting it.

I decided to let Swartz's own voice be the principal one in this story, in an attempt to simulate aspects of a first-person narrative. I have emphasized certain incidents and characters and de-emphasized others, for reasons of thematic relevance and narrative economy. Other ways of telling Swartz's story certainly exist, and I can only hope that any deficiencies in my own approach are offset by its merits. I hope I have contextualized a debate that often seems entirely contemporary, humanized some of its most revered figures, and inspired curious readers to pursue further research on the issues and the individuals covered herein. Any errors and omissions are my responsibility alone.

—Justin Peters
Boston, Massachusetts, July 2015

THE IDEALIST

INTRODUCTION: THE BAD THING

On Friday, January 4, 2013, Aaron Swartz awoke in an excellent mood. He brought his girlfriend, Taren Stinebrickner-Kauffman, up to the roof of their apartment building in Brooklyn, New York. Under overcast skies, he turned to her and said, unprompted, "This is going to be a great year."[1]

The statement was wildly optimistic. Two years earlier, Swartz had been caught using the computer network at the Massachusetts Institute of Technology to download millions of scholarly journal articles from the online database JSTOR. While downloading academic papers is not in itself against the law, doing so in bulk without explicit authorization is—or, at least, that was what the US Attorney's Office in Boston had claimed. Swartz was arrested and indicted on felony computer-crime charges that carried a maximum penalty of ninety-five years in prison and more than $3 million in fines.

Swartz ran no risk of receiving the top sentence. But his friends and family did not think he deserved any prison time at all. When his former girlfriend Quinn Norton met with the prosecutors in April 2011, she marveled at their stubborn insistence that excessive downloading was a serious crime. "I told them this case was ridiculous,

I told [lead prosecutor Stephen Heymann] not to do this," Norton wrote of the meeting. "They listened in silence."[2]

The prosecutors, as prosecutors tend to do, were using the threat of a long sentence to pressure Swartz into pleading guilty. But they were adamant that any plea bargain would have to include some jail time. The government initially offered Swartz a relatively lenient deal: plead guilty to a single felony count and serve a few months in a federal prison, followed by a period of supervised release.[3]

But Swartz refused to sign any plea that would send him to prison. He resented that the supervised release program would restrict his access to computers. Nor was he keen to have a felony conviction on his record. Several months after his arrest, Swartz and Norton had visited Washington, DC. While walking by the White House, Swartz got sad. "They don't let felons work there," he said.[4]

Swartz was twenty-six, small and dark, with shaggy black hair that fell below his ears, and an occasional beard that fell somewhere between stubble and scruff. As a boy he had been ashamed of his body, which he felt was "embarrassingly chubby."[5] But time had straightened and sharpened his features, and now he was handsome and barely resembled the awkward young computer prodigy he had once been. "Teenager in a million," the *Sunday Times* had dubbed him in 2001, when he was fourteen, in an article lauding him as a computer programmer whose gifts far exceeded the norm for his age.[6] At age nineteen, his whiz-kid reputation only grew when Reddit, the Internet start-up he had helped to build, was purchased by the publisher Condé Nast for an undisclosed sum said to be somewhere between $10 million and $20 million.[7]

Swartz never told anyone how much money he earned from the sale, though it was clearly a substantial amount. "His reticence became a running joke between us," Norton wrote, "me prying, cajoling, pressuring, and Aaron never giving in."[8] Swartz kept a lot of things to himself. When Swartz first started dating Taren Stinebrickner-Kauffman, he withheld all details of his legal predicament, not even mentioning that he had been arrested. Instead, he referred to the case, euphemistically, as "the bad thing."

The bad thing came into being during the first few days of January 2011, when network engineers at the Massachusetts Institute of Technology discovered a laptop concealed in a basement closet and wired into the campus network. The computer had been programmed to download documents from JSTOR at a rapid rate. The school installed a surveillance camera and, on January 6, photographed a young white male with dark, wavy hair entering the closet to retrieve the laptop. The man attempted to conceal his face behind a bicycle helmet, but he was clearly Aaron Swartz.

This discovery raised far more questions than it answered. Aaron Swartz was famous. He was neither a malicious hacker nor a vandalistic "script kiddie," but rather a well-known programmer and political activist. He was friends with the Internet icons Lawrence Lessig and Tim Berners-Lee. He was a research affiliate at Harvard. His blog was internationally popular. So why was he skulking around an MIT basement siphoning obscure research papers like some tenure-track cat burglar? What were his plans for the nearly 5 million JSTOR documents he had acquired? What in the world was he *thinking*?

Even two years after his arrest, these questions remained unanswered. The prosecutors believed that Swartz intended to post the JSTOR archives for free on the Internet. Since childhood, he had been affiliated with the "free culture" movement, whose members believe that digital networks ought to *remove* barriers to information access, not erect them—that the Internet ought not to be considered a bookstore so much as an infinite library, its contents made available for the benefit of all.

Free culture is rooted in the public domain, a concept that dates to the earliest formalized copyright laws. In America, when a work's copyright expires, that work falls into the public domain, which means that the public is free to use, modify, and distribute it at will, without having to pay a fee or ask permission. In 1790, when the United States passed its first federal copyright statute, the standard copyright term lasted fourteen years, with a fourteen-year renewal period. By the time Swartz first started downloading JSTOR articles, the standard copyright term for new works in America lasted until seventy years after the author's death, and the public domain had correspondingly dwindled.

The existence of the public domain implies that the public has an inalienable stake in works of culture and scholarship—even works that they did not themselves devise. Stories mean little unless they are told and retold, and in the retelling, new meaning accrues to the original tale. The effect a given work has on its audience—and vice versa—is intrinsic to its social value.

Many free culture advocates believe that copyright terms should be reduced and that the public domain should be reinvigorated. "I want to see copyright regulation shrunken back down to where it came from," Swartz wrote in 2002. "These freedoms likely won't be retroactive, and they certainly won't be easy, but we must try to achieve them."[9] Swartz even argued that restricting public access to useful knowledge was immoral, and that the information-rich were obliged to share their bounty with the information-poor.[10]

This sort of rhetoric horrifies many of the publishers, film studios, record companies, software developers, and other culture merchants whose businesses are based on the artificial maintenance of information scarcity. Radical free culture could drive them into bankruptcy. They see unauthorized online file sharing not as an act of liberation, but of larceny; to them, someone like Swartz wasn't an idealist, but a common thief.

"Stealing is stealing, whether you use a computer command or a crowbar and whether you take documents, data or dollars," said US Attorney Carmen M. Ortiz in a press release that announced Swartz's indictment.[11] From her standpoint, American law and custom clearly establish copyright as a property right, and society has derived clear benefits when the laws protecting property are vigorously enforced.

Writing for the online magazine *Guernica*, days after the indictment, Swartz's friends John Summers and George Scialabba ridiculed Ortiz's simplistic moral tautology: "'Stealing is stealing' is phrasemaking designed to confuse the legal and moral distinctions between the kind of cyber-crime everyone should oppose, such as stealing credit card and social security numbers, and efforts, like Aaron's, to make knowledge more accessible to the educated public. Ms. Ortiz, incredibly, asks the public to ignore the motive behind the act."[12]

But Swartz never publicly revealed his motives. He rarely discussed the case at all, even with his closest companions. His silence seemed impelled by shame, an intrinsic emotional solitude, and a reluctance to involve his friends in his troubles lest they, too, incur the government's wrath. As Stinebrickner-Kauffman later said, Swartz generally believed "that he shouldn't rely on anyone else . . . that strength meant standing alone."[13]

Swartz's self-reliance often proved debilitating, to the extent that even the prospect of asking a flight attendant for a glass of water was enough to inspire paroxysmal guilt. "I feel my existence is an imposition on the planet. Not a huge one, perhaps, not a huge one at all, but an imposition nonetheless," he wrote in 2007. "Even among my closest friends, I still feel like something of an imposition, and the slightest shock, the slightest hint that I'm correct, sends me scurrying back into my hole."[14]

Swartz had been in his hole since his arrest, in a sense, and at times it seemed that he would never be able to emerge. His decision to spurn the proffered plea bargain had irked the prosecutors, and since then the two sides had been at loggerheads. As the April 1, 2013, trial date approached, the lead prosecutor, Stephen Heymann, said that he would request a federal-guidelines sentence of at least seven years in prison if he won the case.

But new years represent new beginnings, and at the start of 2013, Swartz saw reason for optimism. He had new lawyers, from Keker & Van Nest, a San Francisco firm that specializes in intellectual property law and white-collar-criminal defense. His lead attorney, Elliot Peters—no relation to the author of this book—was prepared to argue that the authorities had inappropriately searched and seized Swartz's computer and USB drive, and he hoped to convince the judge to bar prosecutors from using any of the evidence they had found therein. If the suppression motions succeeded, then maybe the prosecutors could be induced to offer a better deal. Maybe Swartz would go to trial and win.

"We're going to win, and I'm going to get to work on all the things I care about again," Swartz told his girlfriend on their rooftop that day. That list was long. Swartz had innumerable interests, and he indulged them all—a trait that occasionally exasperated the colleagues and collaborators who struggled to command Swartz's full attention. He was a programmer and a political activist, an aspiring author who had started and abandoned drafts of several books. He had recently become a contributing editor to a small magazine called the *Baffler*. He was beginning to do research on how to reform federal drug policy. He loved the novelist David Foster Wallace. He watched a surprising amount of television—*Louie*, *Boss*, and *The Newsroom* were favorite shows.

In 2006, Swartz penned a "generalist manifesto" in which he urged his readers to elevate their professional ambitions and transcend their self-imposed limitations. Software engineers, for instance, shouldn't settle merely for writing effective computer code; they should think about how they could apply their talents toward more majestic ends. "People are afraid of grandeur; it challenges the status quo. But you shouldn't be," he wrote. "'Look up more' should be your motto; 'Think bigger' your mantra."[15]

He followed his own advice and came to think of himself as an "applied sociologist," actively working to develop rational strategies for making the world a better place. "Saving the world" had long headed Aaron Swartz's bucket list, and he was intent on achieving that goal. In 2011, he even drafted a document titled "How to Save the World, Part 1," in which he identified and analyzed eight methods that could be used to amass power and influence policy, ranging from innovative public-relations messaging ("In a democracy, changing the world usually means changing the public's mind") to tactical legal pressure ("Litigation in general is a powerful activist strategy").[16]

Swartz planned to save the world by making it more effective. He was a voracious reader of business books: the biography of investor Warren Buffett; the biography of Walmart's founder, Sam Walton; anything that might contain insights into organizational behavior and successful management strategies. At the beginning of 2011, in a capsule review of a book about the Toyota Motor Corporation, Swartz wrote that "lean production"—the Toyota management strategy that prioritized perfectly efficient manufacturing techniques—was "undoubtedly the greatest human art form," and he meant it ("with sex running a close second," he clarified).[17]

Swartz was obsessed with systems, optimizing and improving them; with making things work where they hadn't before. "The revolution will be A/B tested," he liked to say, meaning, essentially, that the most effective organizations recognized their successes and their failures, learned from both, and adjusted their tactics accordingly. "The revolution will be A/B tested" was something of a life philosophy for Swartz. In project after project, he probed and tinkered until he had elicited the answers he was seeking, forever iterating his way toward the ideal outcome, the logical solution.

The bad thing was bad, in part, because it resisted logical analysis. Aaron Swartz had no criminal record. His alleged crime was neither violent nor actively malicious. Swartz returned everything he had downloaded, and JSTOR, conscious of its public image, told prosecutors that it did not want to see him go to prison. Yet the US

Attorney's Office in Boston had delivered a thirteen-count felony indictment and pursued the case with what Swartz and his allies considered disproportionate zeal.

Getting caught inside an irrational, inflexible system—one that made no provisions for his own exceptionalism—was one of Swartz's great fears. He had fled from most of the hierarchies he had encountered during his life. High school, college, the business world: he departed all three prematurely, after finding himself unable to accept their partitions and constraints. The Department of Justice was yet another sclerotic institution, best withdrawn from rather than withstood. But a defendant cannot unilaterally withdraw from a criminal indictment, and that was perhaps the worst, most frustrating thing about Swartz's legal situation: he couldn't improve it, and he couldn't escape it.

In 2013, Swartz's friend Seth Schoen wrote about Swartz's boundless faith in reason. According to Schoen, Swartz believed he would "fix the world mainly by carefully explaining it to people."[18] Some people, however, didn't particularly want to have the world explained to them by a dogmatic young man with a flair for polemics. "He's a young guy who likes to flame," the software engineer Dave Winer, one of Swartz's earliest sparring partners, wrote in 2003. "He's treated me like crap for years, and child or not, I'm tired of it, and I'm not taking it anymore."[19] Though many people thought him brilliant, many others just thought him immature.

Swartz could be dramatic and tendentious, prone to provocative overstatement. As a teenager, he questioned the "absurd logic" of laws that banned the distribution and possession of child pornography.[20] In 2006, he argued that music was perhaps objectively improving over time, and that Bach's *Well-Tempered Clavier* might well be inferior to Aimee Mann's 2005 album, *The Forgotten Arm*.[21] He eschewed business suits on principle, calling them "the physical evidence of power distance, the entrenchment of a particular form of inequality."[22]

"I felt like he certainly had sort of the conviction of youth, in the sense that he was convinced that whatever he was doing at any one time—or at least he sort of projected this—he was absolutely convinced that that was the way to go," his friend Wes Felter said. "And the thing is that most people sort of outgrow that, and I don't know if he ever did."[23]

Throughout Swartz's life, simple stimuli routinely elicited complex reactions, and minor aggravations were routinely magnified into moral crises. A pathologically picky eater, Swartz preferred bland, achromatic foods: dry Cheerios, white rice, Pizza Hut's Personal Pan cheese pizzas. ("This reached its extremes at a World Wide Web conference where all the food was white, even the plate it was on," Swartz wrote in 2005. "Tim Berners-Lee later pulled my mother aside to share his concerns about this diet.")[24] He suffered from ulcerative colitis, which partially explains his limited palate. But he also told friends he was a "supertaster," extraordinarily sensitive to flavor, as if his taste buds were constantly moving from a dark room into bright light. His friend Ben Wikler suggested that Swartz was also a "superfeeler," forever oscillating between emotional extremes.[25]

On the morning of Friday, January 11, 2013, Swartz seemed particularly morose, in sharp contrast to his earlier optimism. He was slow to rise from bed and told Stinebrickner-Kauffman that he was going to stay home from his job at the global tech consultancy ThoughtWorks. "I said, 'Well, why don't I stay home with you, and we'll go for a hike? We'll go for a hike today,'" Stinebrickner-Kauffman remembered. "And he said no, that he just really wanted to rest, and he needed to be alone."[26]

When she got to work, Stinebrickner-Kauffman contacted Wikler and organized a dinner party for that evening, in hopes that the gathering might lift Swartz's spirits. Wikler had introduced the two of them at the end of 2010, toward the end of Swartz's relationship with Quinn Norton. "In the first eight weeks that I started dating him, he quit two jobs, broke up with his ex-girlfriend—Quinn—moved from Cambridge to New York, and was indicted," Stinebrickner-Kauffman recalled. "So it was kind of a big time in his life."[27]

Despite these adverse circumstances, a romance flourished between the two. Early in their courtship, Swartz joined Stinebrickner-Kauffman and two mutual friends on a strawberry-picking excursion, much to his evident displeasure. "Strawberry picking wasn't his idea of a productive way to spend the afternoon," one of the holidaymakers, Nathan Woodhull, recalled, in a significant understatement. "It was very much Taren's idea. And I remember joking to her, later, that he must have really liked her."[28]

In the spring of 2012, they began to cohabitate, sharing a small apartment near Prospect Park in the Crown Heights neighborhood of Brooklyn. "We would joke about how the two of us would really confuse the census," Stinebrickner-Kauffman said.[29] "We could be reasonably recorded as two unmarried high school dropouts living together in a one-room studio." As their relationship progressed, Swartz gradually began to bridge the emotional distance that had long set him apart from the world. "Early in our relationship he didn't want to feel like he was dependent on me in any way," Stinebrickner-Kauffman remembered. "And I think that changed a lot. I think that he saw us as a pair that could compensate for each other's strengths and weaknesses."[30] As 2012 drew to a close, their love was stronger than ever before.

On the evening of January 11, 2013, Stinebrickner-Kauffman stopped at Ben Wikler's apartment on her way home from work. As she played with Wikler's newborn son, she mentioned that Swartz had told her that he might consider getting married after his case was resolved.[31] Swartz had long scorned the idea of marriage, but he was beginning to reconsider his opposition. If a reversal like that was possible, then anything was possible.

But less than two miles away, in a small and dark studio apartment, Aaron Swartz was already dead by his own hand.

———————

A plan, an arrest, a prolonged indictment, a resolute prosecutor, a suicide—these are the undisputed facts of Aaron Swartz's case. They are outnumbered by the questions that Swartz's family, friends, and supporters began asking soon after his death. Why had the US attorneys been so intent on sending Swartz to prison? Why did Swartz choose to hang himself rather than go to trial or accept a plea bargain? How did academic research papers come to be considered private property, protected by law? How did accessing that material without explicit authorization come to be considered a federal crime? These questions remain vexatious even now, almost exactly three years later. This book is an attempt to address them.

The story of Swartz's life and the circumstances of his death are recent inflection points in a contentious debate over the means by which information circulates in society and the laws and technologies that speed or delay its progress. Aaron Swartz has become an avatar for a movement, his actions and presumed intentions an argument that the government ought to pass laws that promote, rather than inhibit, the digital dissemination of knowledge.

This debate is not new. The epidemiology of information has been a public-policy concern since the preliterate era, when news and legend spread orally, allowing fealty and rebellion to slowly infect what I'll call the cultural brain. The cultural brain is the zeitgeist, the ether, the intangible repository of communal interests. The cultural brain is *the conversation*. Its changing makeup is a function of technological progress. The oral tradition, the printing press, the railroad, the telephone, the radio, the television, the Internet—all of these innovations opened channels for what otherwise might have remained stray ideas, and gave those ideas velocity and direction. They are the mechanisms by which an entire society can come to consider and discuss the same ideas and events. Lawmakers have never been quite sure whether to feed or starve the cultural brain.

As communications technologies have advanced, this conflict has intensified.

In his 1987 book, *The Media Lab*, the entrepreneur and futurist Stewart Brand memorably asserted that "information wants to be free": that it is effectively impossible to restrict the flow of (and artificially maintain high prices for) data in a world rife with photocopiers, tape decks, instant cameras, digital networks, and other such disseminative tools.[32] Brand was the founder of the *Whole Earth Catalog*, which, in the 1960s and 1970s, brought long-haired shoppers a message of conscientious consumption. In the 1980s, Brand became interested in digital networks. He thought that, like the tools he had featured in the *Catalog*, these networks had the potential to bridge cultural chasms and empower their users to transform society.

"Information wants to be free," with its air of casual inevitability and hazy, imprecise idealism, is less an insight than a bumper sticker, easy to chant, easier to dismiss. Perhaps this is why the phrase resounded. It was pithy, unapologetic. Of course information wants to be free: the more people who are familiar with a particular piece of information, the more relevant that information becomes. But fewer people remember what Brand wrote immediately thereafter:

> Information wants to be free because it has become so cheap to distribute, copy, and recombine—too cheap to meter. It wants to be expensive because it can be immeasurably valuable to the recipient. That tension will not go away. It leads to endless wrenching debate about price, copyright, "intellectual property," and the moral rightness of casual distribution, because each round of new devices makes the tension worse, not better.[33]

Brand's paradox describes the state in which we find ourselves today, and neatly summarizes the ideologies that have driven the last half century's worth of data activism. While the members of the copyright bar drafted laws ensuring that information would remain expensive, engineers and futurists constructed a medium that would help set that information free. Since the earliest days of digital computing, idealists have envisioned the machines as the foundation of an infinite library that would offer unfettered access to the fruits of human knowledge and creative production. The "library of the future" would be intuitively organized and universally accessible; it would be responsive, personalized, and intelligent; it would belong to everyone and benefit everyone.

But this dream has consistently been deferred, perhaps because it is and has always been wildly unrealistic. The informaticists Rob Kling and Roberta Lamb characterize computing technology as the "centerpiece of seductive dreams," a license for digital soothsayers to emphasize computing's utopian potential while ignoring the sociopolitical realities that might thwart ideal outcomes.[34] Theoretically, computers and the Internet can be used to promote congruence, tolerance, and understanding. In the real world, however, forward progress will always be slowed by social and political friction, often generated by those who do not think these goals represent progress at all. Internet enthusiasts often presume that the network inevitably leads to *yes*, even though the world has always, always, always been defined by *no*.

For more than a century, Congress and America's content industries have been of one mind on copyright policy, consistently advocating for longer terms and harsher penalties for scofflaws, nominally in order to empower individual creators, but also to maintain the commercial viability of the companies that publish and distribute those works. Most major new advances in communications have been greeted with a barrage of rhetoric and legislation meant to inhibit the new technologies for the benefit of existing ones.

Meanwhile, unauthorized access to intellectual property has consistently been characterized as piracy, the perpetrators as bespectacled Blackbeards intent on ravaging the cultural commonweal. Statute and morality have been conflated by copyright holders seeking to extend their monopolies and by legislators looking to validate their decisions to pass laws that do so.

Today, the Internet is an information smorgasbord, an endless buffet of data sets, research papers, essays, news articles, instruction manuals, annual reports, and myriad other materials that the curious soul can use for edification and self-improvement. But much of this content is inaccessible, consigned to subscription databases or hidden behind paywalls or subject to other impediments. The Internet has simultaneously shrunk and expanded the world with its hyperlinked data and its decentralized, disintermediated communications. And it has consistently confused and frustrated legislators and capitalists who don't understand why so many users are so insistent on flouting The Rules.

Aaron Swartz broke The Rules, consistently and creatively. He did not always explain why, but he broke them, and now his death has led many people to question those rules and wonder why they seem so incongruous with the realities of the digital era, in which information has become untethered to physical formats and it takes only moments for news and legend to permeate the cultural brain.

Three years after Aaron Swartz died, his story is still on many people's minds. A large street-art mural of his face, set next to the words RIP AARON SWARTZ, adorns the side of a building in Brooklyn. The filmmaker Brian Knappenberger turned Swartz's life story into the documentary *The Internet's Own Boy*, which played at the Sundance Film Festival and was released in theaters during the summer of 2014. Every year around his birthday, Swartz's friends and admirers worldwide organize a series of weekend-long "hackathons" intended to stimulate the sorts of social projects Swartz cherished. A cottage industry of programmers and political activists has arisen to ensure that Swartz's legacy will survive.

Of course, to tell the story of Aaron Swartz requires more than *just* the story of Aaron Swartz—the difficult, immoderate men and women who preceded him as caretakers of the cultural brain must also be included. They share a spiritual kinship with Swartz, confident, as he was, not only in the righteousness of their causes, but that they themselves were destined to advance them. Their lives, passions, accomplishments, and failures provide a necessary context for Swartz's life and death. His story is inextricably twined with theirs and cannot adequately be understood without realizing that his predecessors' lives and works, taken together, created the system that produced, defined, and destroyed him. Aaron Swartz wanted to save the world. But the world was never going to let Aaron Swartz save himself.

1

NOAH WEBSTER AND THE MOVEMENT
FOR COPYRIGHT IN AMERICA

A book is intrinsically different from a boat. Both books and boats
have definite physical properties: you can see them, touch them,
occasionally smell them, and even taste them, if you're into that
sort of thing. But a good boat benefits no one but its owner. Society
at large derives no value from your boat, unless you are a noble
soul who throws come-one-come-all pontoon parties at your summer place on Green Lake. Society can, however, profit greatly from
a good book—or, more precisely, from the good ideas contained
therein.

While ideas can improve society, they can also imperil it. The
history of civilization is, among other things, the history of powerful
people suppressing heterodox ideas, usually by executing their bear-
ers. Chopping blocks and crucifixes are conclusive, but also messy,
and they sometimes make martyrs of their victims. Martyrs inspire
movements; movements are bad for business. The earliest copyright
policies were, in a sense, slightly more humane methods of extin-
guishing dissent.

Copyright first became a relevant policy question with the European development of movable type circa 1450, when the printer Johannes Gutenberg unveiled his press in the city of Mainz.[1] Before then, book copying was an arduous task, done by hand, and governments did not have to fear the rapid scriptorial dissemination of seditious or heretical ideas. Most people were illiterate; culture traveled orally. The printing press made the advance of a general literary culture possible for the first time in human history. Movable type was an accelerant, the printed book a preservative.

By 1500, Gutenberg's invention had traversed the Continent, and the governments of Europe saw that this new technology could be used to spread troublesome sentiments, which, if widely felt, might destabilize their regimes. Various rulers and their ministers gave this matter some thought, presumably realized that it would be impractical to forcibly lobotomize the reading public, and instead decided that the answer was to regulate what the press could print, and who could do the printing.

In England, this regulation was accomplished by the royal charter, in 1557, of the Stationers' Company, a guild comprising the licensed printers in the realm.[2] The ninety-seven men named in the Stationers' charter got a pretty good deal: a near-total monopoly on publishing in England as well as the power to fine and jail illicit printers and seize and destroy their output. By this inky circumscription, Queen Mary explicitly intended to disempower the many "scandalous malicious schismatical and heretical persons" who persisted in publishing works that insulted or criticized church and crown. The Stationers were simultaneously printers and censors.

This first foray toward copyright was meant to benefit neither authors nor the public, but the state. The Stationers' Charter had nothing whatsoever to do with an author's right to make a profit, and everything to do with consolidating royal control over printed matter. The publisher held perpetual copyright. The author got little but the diminishing marginal satisfaction of seeing his works in circulation. Once those works were in print, as the essayist Augustine

Birrell observed in 1899, "the British author usually disappeared, or if he did reappear, it was in the pillory."[3]

The Glorious Revolution of 1688 upended this relationship, and many other things, by establishing Parliament as the fountainhead of governance in England. The stature of the Stationers and the Crown's control over printed material slowly eroded as the seventeenth century approached its end, culminating in Parliament's failure, in 1695, to renew the licensing act granting the Stationers their monopoly and powers of proscription. Faced with increased competition, declining profits, a market newly glutted with rival editions of the same books, and an unpleasant sense of existential dispossession, the Stationers pioneered what would soon become a standard industry tactic: they tried to get the copyright laws changed in their favor.

They did not immediately succeed. The Stationers lobbied for the restoration of their monopoly in 1703, 1704, and 1706, and each time found Parliament unmoved by their argument.[4] They had better luck in 1707, when they suggested that a new copyright law would help not just printers and booksellers, but the authors of England and their families, who had ostensibly been driven to poverty and ruin by literary deregulation and the ensuing market chaos. Augustine Birrell later observed that the Stationers' rhetoric about the plight of the British author "has no foundation in the facts of literary life in England."[5] Nevertheless, their claim was apparently convincing, and in 1710 Parliament passed a law that restored at least *some* order to the world of English literature.

The Statute of Anne, as the law was called, was the first modern copyright statute and subsequently became the model for future copyright legislation passed in the United Kingdom and the United States. Named after the reigning British monarch, the statute rejected the notion that perpetual copyright was a natural law akin to those of gravity, attraction, or optics. Instead, it established a fixed copyright term of fourteen years from the date of a new book's initial publication, renewable once. Moreover, the act decreed that the copyright belonged to the *author*, as opposed to the publisher or printer. In so doing, the Statute of Anne indicated that copyright existed not as a censorship tool, but as an incentive to produce—an act of public policy, not an assertion of moral rights. Authors weren't the only parties empowered: the law was deemed "An Act for the Encouragement of Learning," meant to increase the production of "useful Books" from which the entire kingdom might benefit. Copyright had been reframed.

Though the law was disliked and disputed by many, it lasted, more or less, for over a century. For the British author, the Statute of Anne did far more good than harm. But the Statute of Anne had no direct effect on the American author, at least not at first. The law did not apply in the colonies, and besides, at the time of its passage there weren't any American authors to speak of. The printing trade there was nascent; there was scant indigenous literary culture. As of 1775, the thirteen colonies had a mere thirty-seven newspapers, most of them based in major cities.[6] The most common books were Bibles and other religious texts. Despite its occasional aphorists, theologians, and pamphleteers, colonial America was an inarticulate place, and its residents were not expected to have their own ideas.[7]

They had them anyway, and after the colonies won their independence in the Revolutionary War, Americans began to compile these stray ideas into a coherent national thesis statement. Fundamental was the notion of free expression: that citizens of the republic were allowed to say what they wanted, whether or not they had anything to say. Eventually, lawmakers came to see the merit in encouraging the expression of original and articulate *American* ideas, as opposed to reheated British ones. This intellectual nationalism came to constitute the most compelling argument for an omnibus copyright law. That argument was taken up most vigorously by a loud and brilliant young man named Noah Webster.

WEBSTER was born in 1758, at once ten years too late and 220 years too soon. He would have made an outstanding Framer, if only he'd been old enough, what with his patriotic zeal, his rhetorical gifts, and his lifelong interest in telling his fellow citizens what was best for them. He would probably have thrived on the Internet, too, with its taste for disputatious pedantry, and its core constituency of strident, insecure nerds. But Webster missed his moment, and though he eventually found a niche, he nevertheless spent much of his life chasing the acclaim that could well have been his but for bad timing.

The fourth child of a resoundingly normal Connecticut farmer, Webster matriculated at Yale in 1774, at age sixteen, and did not fight in the Revolutionary War. (Later in life he would rather pathetically offer an anecdote about how, as a volunteer militiaman in 1777, he had "shouldered the best musket he could find" and "marched up the Hudson" to help fight the Battle of Saratoga. Left unsaid was that the battle had already ended by the time he arrived.)[8] At Yale, he studied classics and divinity, ate "cabbage, turnip and dandelion greens with plenty of cider passed round in pewter cans,"[9] and delivered one of the class orations upon graduation in 1778, on the theme of natural philosophy and the merits of empirical inquiry. Webster "was committed to bringing order to the world through his intellectual labors," wrote the biographer Joshua Kendall, "though he hadn't yet figured out exactly what those labors might be."[10]

He returned home to a destitute father and a cloudy future. The war had sapped the elder Webster's savings, leaving him unable to support his son's continued education. His father handed him a devalued wartime banknote and, effectively, told his son that he must now become a man. "Take this," Noah Sr. said, "you must now seek your living; I can do no more for you."[11]

Webster sulked briefly, and then, like countless thwarted philosophers before and since, decided that, for lack of better options, he would become a teacher.[12] So he opened a succession of schools, first in Connecticut, then in New York, where he taught wealthy children how to read, write, and—occasionally—sing. The standard spelling text at the time was an arid volume called *A New Guide to the English Tongue*, written by an English schoolmaster named Thomas Dilworth. Webster soon came to despise both the book and, rather unfairly, its deceased author.

For one thing, Webster thought it unpatriotic to teach young Americans to read and write using a textbook that glorified the nation's former colonial overlords. "[Dilworth] has twelve or fifteen pages devoted to names of English, Scotch and Irish towns and boroughs. Whatever purpose these may have served in Great-Britain, they certainly are useless in America," Webster sniped.[13] "And yet ten thousands of these books are annually reprinted and find rapid sale, when one half of the work is totally useless and the other half defective and erroneous." So Webster decided to write his own book of grammatical instruction, one that was neither useless nor defective, one that was deliberately tuned for American ears.

Webster believed that proper pronunciation could help save the United States—the formation of which, in the early 1780s, still struck some people as a reckless mistake. The Revolutionary War had been won, the British subdued, leaving behind a loose alliance of self-governing states united by geographical proximity, the memory of battlefield camaraderie, and a common language, sort of. Though English was the new nation's mother tongue, its various regional dialects threatened to perpetuate the sorts of class divisions to which the republic was ostensibly opposed. While "all men are created equal" is nice in theory, tacit class divisions would always exist in America as long as educated men spoke the King's English and drawling yokels persisted in saying *sparrowgrass* for *asparagus*, *ax* for *ask*, *chimbley* for *chimney*. A nation divided over the proper pronunciation of *asparagus* could not stand.

The idea that a country's grammar is central to its collective identity is more than just an English teacher's aspirational delusion. The words and constructs we use to define and articulate the world around us affect the way we understand it. Webster wrote of "the influence of a national language on national opinions," and his ambition to "detach this country as much as possible from its dependence on the parent country."[14] How could the United States ever be truly independent if its linguistic customs were wholly indistinguishable from those of its former masters?

"America must be as independent in *literature* as she is in *politics*, as famous for *arts* as for *arms*, & it is not impossible, but a person of my youth may have some influence in exciting a spirit of literary industry," Webster wrote in 1783.[15] Webster had not been one of the heroes of the American Revolution. But he was determined to win glory in the nation's battle for cultural independence.

Webster published his spelling textbook in 1783. The official title was *A Grammatical Institute of the English Language*, but since that name was both long and terrible, most people just called it the "blue-backed speller," after the color of its cover.[16] The book emphasized instruction on the syllabic level. It was filled with stories and sentences designed to appeal to the curious American child. The *Institute*, he believed, would help demolish the possibility that an American might be judged on the way he spoke rather than on what he said. With luck, it might also make Webster rich and famous. For all those things to happen, the book had to be ubiquitous. And for *that* to happen, Webster needed to do two things: secure endorsements from the country's most prominent men and put the book under copyright.

Thus the story of copyright in America begins, more or less, with Noah Webster, the self-promoting, painfully insecure lexicographer who was determined to profit from his peculiar greatness. Webster was a lobbyist and a proselytizer, a prig, a pedant, a prodigy. "His mind was not subtle or graceful; he had not the faculty of creating, nor, so far as I can discover, of appreciating literature," wrote his first biographer, Horace Scudder.[17] For most of Webster's life, according to the cultural historian Tim Cassedy, "the overwhelming majority of what he wrote was met with jeers and boos in newspapers and magazines throughout the United States."[18] Webster's passions often made him a pariah, and, while it would be incorrect to say that he did not care about being vilified—as one of the most insecure great men of his era, he did, very much—he didn't let the public resistance prevent him from pursuing his goals.

While the concept of copyright was not novel when Webster first took it up, and while he was certainly not its only American advocate, he is nevertheless an important figure, not just for all that he did in the service of literary property, but for how he did it, and for who he was. If the story of copyright in America begins with Noah Webster, then the following chapters have often belonged to people *like* Noah Webster: earnest, linear young men and women who resolved to harangue a largely indifferent populace into caring about intellectual property and its sociopolitical implications. Their arguments were occasionally self-serving, and almost always framed in moral and patriotic terms: strong copyright laws, they maintained, were good for authors, for publishers, and for the welfare of Christian America. "Information is fatal to despotism," Noah Webster wrote in 1788, and a literate, informed public was an unchained public.[19] Webster was among the first in a line of Americans who came to believe that good information, widely disseminated, would make the nation stronger and more free. And as much as anyone else, Webster can also take credit for teaching Americans that the most valuable material ought to be locked away.

ALTHOUGH Noah Webster weathered a lot of criticism in his time, even his fiercest foes would have acknowledged his diligence and ambition. By the time he was only twenty-seven years old, Webster had already written and published a series of grammar textbooks and a collection of political essays—an impressive feat in a time when "author" was about as sensible a career ambition as "astronaut."

While it would be inaccurate to say that Americans in the immediate postwar period had no interest in books—to the contrary, the Revolution made new readers out of common people who had never before had any immediate incentive to participate in civic life—the young nation was not situated to cultivate its own writers. The publishing historian William Charvat has written that, even as late as 1820, there were no full-time literary authors in America. The shortage is at least partially attributable to the republic's rickety copyright and printing infrastructure.[20]

At the time that Webster published his speller, American authors generally assumed the financial risks for their works, as publishers "lived on such a narrow margin that not many had a life of more than a few years," wrote Charvat.[21] Impoverished authors could solicit prepublication pledges from wealthy "subscribers" who agreed to prepay for copies of the forthcoming work—an early version of today's online crowdfunding platforms, and one just as tiresome as its modern-day incarnation. But early America had few wealthy and willing subscribers; as Webster's friend Oliver Wolcott Jr. put it years later, "It is in vain to reason with the greatest part of mankind, if they have to pay Ten Dollars, in consequence of being convinced."[22] So, for many aspiring authors, self-financing remained the best option, even though it was objectively a bad one. The scholar James N. Green described how post-Revolutionary American writers who underwrote their own works "had hopes of making a profit and perhaps even a living from their books, but most were disappointed, and some—like David Ramsay, who financed the printing and binding of his two histories of the American Revolution—lost heavily on their investments."[23]

The structural inefficiencies of the book business in early America further impeded authorial success. Printing was a shambolic trade, and writers could not always ensure the fidelity of the final product to their original manuscripts or prevent printers from issuing unauthorized editions of the same. In a 1783 letter to the president of the Continental Congress, the poet Joel Barlow noted that his American compeer John Trumbull was unlucky enough to see his epic poem *McFingal* "reprinted in an incorrect, cheap edition" and thus "suffers in his reputation by having his work appear under the disadvantages of typographical errors, a bad paper, a mean letter & an uncooth [*sic*] page, all which were necessary to the printer in order to catch the Vulgar by a low price."[24] Robust copyright laws, Barlow believed, would be the American author's best weapon against the Vulgar.

Barlow's letter inspired the Continental Congress to issue a resolution, in May of 1783, encouraging the various states to draft and pass their own copyright laws. (In this pre-constitutional era, the United States was organized under the Articles of Confederation, which limited federal powers and asserted state sovereignty.) But most states were none too quick to respond. If the rise of statutory copyright in England had been contentious because it imposed a finite term of control on works that had once enjoyed perpetual protection, it was controversial in America because it asserted ownership rights where *none* had existed before.

The copyright scholar William Patry has suggested that framing intellectual property as a "right" allows owners to "launder self-interest as enlightened conduct" and turn policy discussions into moral crusades.[25] It would be more fair and equitable, in Patry's opinion, to characterize copyright as a set of mutable social relationships between creators, the public, and the state, and to legislate and mediate with all parties' interests equally in mind. The public benefit in cheap books, for example, should not always and automatically be subordinated to the private benefit in dear ones.

In England, statutory copyright had come about after intense lobbying by printers. The movement for copyright in America was led by authors. Absent from both parties' self-interested arguments was a clear explanation of how these laws would affect readers. While there exists no comprehensive statistical documentation of literacy rates in eighteenth-century America, anecdotal evidence suggests that the nation contained plenty of readers. As Julie Kay Hedgepeth Williams wrote in her doctoral dissertation on the printed word in early America, "Colonial Americans craved and used the printed word. They depended on it, bargained for it, swapped and loaned and made gift of it. They admired it, revered it, and were in awe of its power."[26] And they might well have looked askance at laws that would perforce affect its circulation.

To obtain nationwide copyright laws, Webster and his comrades had to frame copyright as a truly American concept, a means of creating and stabilizing a literary middle class by allowing writers to control and profit from the sale of their works. To ensure that his books would be copyrighted, Webster decided to take responsibility for his own destiny and lobby almost every single legislature on his own behalf.

WEBSTER'S first forays into lobbying were comically inept. In the fall of 1782, before he had even published his speller, Webster visited New Jersey and Pennsylvania in hopes of "obtaining a law for securing to authors the copy-right of their publications."[27] Unfortunately, both state legislatures were out of session. He next stopped in Connecticut, where he petitioned "for a law to secure to me the copy-right of my proposed book."[28] While the Connecticut state legislature was actually in session—a small victory—it was too busy to entertain Webster's proposal. In January 1783, the undaunted schoolmaster again attempted to petition the Connecticut state legislature for copyright in his own work. The state did him one better and passed the first general copyright law in the nation.[29]

Webster eventually decided to expand his lobbying efforts, and in 1785 he embarked on an extended lecture tour during which he made the case for copyright—and for himself—to legislators and civic leaders in most of the Southern states. His strategy was simple. He would enter a city armed with letters of introduction to its most prominent citizens. If, as sometimes happened, he entered a town wholly unrecommended, he would find other ways of announcing his presence. At many of his stops, Webster found time to deliver a series of dull lectures on the English language; although Webster knew that these presentations "were not very interesting to a popular audience"—an understatement—the sort of people who *did* attend were apt to act as allies down the line.[30] "Obtain permission to read Lectures in the State House by vote of the House of Delegates," Webster wrote in his diary on January 5, 1786. "Begin to read—to about 30 respectable people."[31]

The contemporaneous term *tuft-hunter* comes to mind when considering Noah Webster's early years, and his gift for courting "respectable people" in the service of his own weird ambition. His diary from this era is filled with brief descriptions of governors met, professors visited, generals imposed upon, all in pursuit of references, testimonials, and associative glory. Though it appears that Webster was usually received with great courtesy, one wonders what his putative patrons actually thought of the overbearing young man who was so eager to share his theories and opinions on the national character and the English language, and how each influenced the other.

For Noah Webster did not come by charm naturally. His overweening self-regard led critics to dub him a "conceited coxcomb," an "inflated fop." A girl with whom the young Webster was friendly once compared his literary talents disfavorably to that of her horse. ("In conversation he is even duller than in writing, if that is possible," she added.)[32] He styled himself "Noah Webster, jun. Esq." and more than lived up to that title's implied pomposity. Even his friends acknowledged that Webster's unabashed egotism served to "prevent his hearers [from] receiving the satisfaction which might otherwise have been derived from many ingenious observations."[33] Like a truffle dipped in ipecac, a gift wrapped with poison ivy, Webster's good ideas were often ruined by their repellent packaging.

In November 1785, for example, Webster set out from Baltimore toward northern Virginia to visit the great George Washington.[34] Though Washington probably wasn't expecting him, that was of no real concern; in the immediate post-Revolutionary period, Mount Vernon had become less a country estate than a tourist hostel, besieged by presumptuous young men seeking close-up glimpses of greatness. No longer a military commander, but not yet president, Washington had ample time to greet the gawkers and indulge the conceit that, in America, any freeborn nobody could ride right up to the home of the country's most prominent citizen, exchange greetings, and come away convinced that even the most eminent man is still just a man.

Mount Vernon was a way station on Webster's trip to Richmond, Virginia, where he hoped to lobby the state's leaders on behalf of copyright protections. Webster hoped Washington would write him some letters of introduction, preferably ones that did not stint on praise for Webster and his work. Although Washington had previously declined an opportunity to endorse the *Institute*, a letter of introduction from George Washington was a valuable commodity, and Webster must have felt there was no harm in trying again.

But, true to form, Webster soon got in his own way upon arrival. Over dinner, Washington mentioned that he had need of a personal secretary and tutor for his wife's grandchildren and was hoping to hire a gentleman from Scotland.[35] Webster was appalled by this news. A Scotsman at Mount Vernon? A subject of the Crown molding the Custis children's minds? In his autobiography, Webster recalled the moment when he asked Washington to consider "what European nations would think of this country if, after the exhibition of great talents and achievements in the war for independence, having obtained our object, we should send to Europe for secretaries, and men to teach the first rudiments of learning":

> The question was well received; and the General instantly replied, "What shall I do? There is no person here to be obtained for my

purposes." N.W. replied he believed any northern college could furnish a person who would answer his wishes. Here the conversation at table ceased.[36]

More than two hundred and thirty years later, the awkward silence still rings. In Webster's telling, the general was suitably chastised and quickly saw the merit in the grammarian's argument. But Washington's true feelings can perhaps be divined from the tepid recommendation he handed Webster upon the young man's departure. Washington wrote that "there are very honorable testimonials" to the *Grammatical Institute*'s "excellence & usefulness," but declined to offer any testimonials of his own. "The work," he wrote, "must speak for itself," or at least Webster would have to speak for it, which he was always happy to do.

So, one after another—though not solely due to Webster's exertions—the various states passed their own copyright laws. By the time Webster concluded his lecture tour and returned to Connecticut in May 1786, every state but Delaware had a copyright statute on its books. Then, in 1790, with the Articles of Confederation a receding memory and a new federal government organized under the principles of the Constitution and the leadership of President George Washington, the first Congress passed a federal copyright law that superseded the various state statutes. ("And Congress too, with powers extending / Farther than patching work and mending; / Have now begun, in freedom's cause, / A code of energetic laws," wrote Webster, who was moved to verse on the occasion of Congress's convention.)[37]

The Copyright Act of 1790, which was based on the Statute of Anne, provided for an initial term of fourteen years, renewable once. "Based on the Statute of Anne" is an understatement; as the legal historian Oren Bracha recently noted, "The similarity is felt on every level, including structure, legal technicalities, and specific text."[38]

Like the Statute of Anne, the American bill was framed as "an Act for the encouragement of learning," a matter of public policy instead of as a natural right. Maps, charts, and books were protected under the act, with the term *books* liberally interpreted to encompass printed works ranging from from catalogs to calendars.[39] Authors had to register their works formally with their local district court before copyright was conferred. Many authors did not bother to do so.

By today's standards, the Copyright Act of 1790 was a paltry grant. Even by the standards of its own time, wrote William Patry, the act was flawed. The registration requirements were onerous, and "practical or commercially useful books, such as works of instruction, textbooks, manuals, geographical atlases, and commercial directories, comprised the bulk of the registrations."[40] Still, the law allowed Webster and his peers to assert their rights to profit from and control their works, and it was the first federal step toward normalizing the concept of intellectual property in America.

What the Copyright Act did *not* do was launch a new literary culture in the United States. According to the copyright historian Meredith L. McGill, "The growth of American print culture did not depend on copyright. In the early republic, a national print culture was established less through the sale of books than through the distribution of uncopyrighted newspapers, magazines, tracts, and pamphlets through the United States mail."[41] Much more meaningful for American cultural flowering was the Post Office Act of 1792, which set favorable postage rates for newspapers and magazines, thus easing the dissemination of news. (The historian Richard R. John observed that the Post Office Act also helped build America's stagecoach infrastructure, since stage operators received lucrative government contracts to carry the mail.)[42]

The newspaper remained the nation's primary vessel for cultural discussion. Most American citizens had "a strong inclination to acquire . . . the means of knowledge," Webster wrote in 1793. "Of all these means of knowledge, Newspapers are the most eagerly sought after and the most generally diffused."[43] In 1790, America had only ninety-two newspapers. By 1800, there were 235.[44] In 1793, Webster became founding editor of *The American Minerva*, the first daily newspaper in New York City, and issued a four-page edition six days per week.[45] (The *Minerva* was eventually rechristened the *Commercial Advertiser*.) Many of the era's newspapers were vituperative and partisan, and the *Minerva* was no exception. Though he claimed to reject factionalism, Webster the editor "was soon unable to see morality in any political position but his own," remarked the historian Gary R. Coll.[46] Webster's editorial tenure was tumultuous, and the newspaper business was not lucrative. He resigned in 1798, worn-out by his labors and discouraged by the antagonistic turn the national discourse had taken. Five years later, in a letter to Oliver Wolcott Jr., he acknowledged, "I regret now that I have devoted my time & the vigor of life to a business so precarious, but regret is useless."[47]

In the early years of the nineteenth century, Webster was filled with regrets, mostly pertaining to his own credulous youthful populism. During the Revolution and immediately thereafter, Webster imagined that the former colonists were creating a level society, free from artificial hierarchies and class divisions. His grammatical reforms had been offered in service of that great cause. But as Webster aged, he began to realize that, while information might be fatal to despotism, an informed, egalitarian society is not necessarily an enlightened one.

Webster was a John Adams Federalist, a proponent of a strong national government, preferably governed by citizens just like himself: sensible, rational men who settled their disagreements via polemics and correspondence. He watched in horror as Americans took to political parties and began to elevate factional concerns over the public welfare. He worried that this discord was the result of a

French plot to divide the nation, and his letters from this time are rich with suspicions of conspiracy. "The people of America, Sir, will not be *tricked* out of their independence in the mean, dastardly manner by which the French extend their conquests," Webster wrote to Joel Barlow in 1798, in response to a letter in which Barlow had the temerity to suggest that perhaps the French *weren't* actively plotting America's demise.[48]

The election of Thomas Jefferson—former minister to France—to the presidency in 1800 just fed Webster's fears. To Webster and other Federalists, Jefferson was a fool and a villain—and, worse, a populist. In a letter to the physician Benjamin Rush, Webster gloomily noted, "We have, by our constitutions of government and the preposterous use made of the doctrines of *equality*, stripped *old* men of their dignity and *wise* men of their influence, and long, long are we to feel the mischievous effects of our modern policy."[49]

Webster's ideal America had not come to pass. "All [Webster's] measures are exploded, his predictions have proved false, not a single sentiment of his has become fashionable," jeered the journalist William Cobbett.[50] Not only was Webster unfashionable, he was poor. Despite the continued popularity of his speller—by 1804, the book was selling at a rate of two hundred thousand copies per year—Webster lived perennially on the brink of insolvency, usually thanks to his own financial incompetence. "My income barely supports my family, and I want five hundred dollars' worth of books from England which I cannot obtain here, and which I cannot afford to purchase," he wrote to Barlow in 1807.[51] Webster eventually condemned the primacy of partisan political chatter in the nation's cultural conversation. Little room was left for more scholarly debates, he complained to Oliver Wolcott Jr.: "vast sums of money [are] expended in donations to support a party or a newspaper—when not a cent can be obtained for very valuable purposes."[52] The Copyright Act of 1790 had portended the arrival of a republic of letters. But its citizens seemed primarily interested in the poison pen.

Under these circumstances, Noah Webster began the project that would be his legacy: the *American Dictionary of the English Language*. His plan was madly presumptuous. Webster proposed to improve upon the work of Samuel Johnson, the celebrated British lexicographer and coffee wit whose own *Dictionary of the English Language* had been well beloved since its first publication in 1755. Just as he had publicly criticized Dilworth when he was assembling the blue-backed speller, Webster justified his project by repeatedly asserting that Johnson's *Dictionary* was fatally flawed. In October 1807, he assured the historian David Ramsay that "not a single page of Johnson's Dictionary is correct: every page required amendment or admits of material improvement."[53] (What's more, Johnson's lexicon contained a shocking amount of "vulgar words and offensive ribalry.")[54] Webster's work would be a work of scholarship and reference; a learned bulwark against ignoble populism.

He spent two decades preparing the work, much longer than he originally anticipated. ("His original research plans had to be extended because of his lack of knowledge of the origin of words," historian K. Alan Snyder drolly observed.)[55] Webster labored from inside a doughnut-shaped table on which rested "dictionaries and grammars of all attainable languages."[56] During the years he spent sequestered in etymological toil, America began to emerge.

The American postal system continued to expand, swelling from 903 post offices in 1800 to 2,300 in 1810.[57] As went the mail, so went the media; the number of American newspapers grew with the postal system, and early-nineteenth-century newspaper readers "treasured each issue as a key to a wider world and a solvent of their isolation," in the words of historian Andie Tucher.[58] ("I am often asked, what progress I have made in the compilation of my proposed Dictionary; and when in all probability it will be completed. To these questions I am not able to give precise answers," Webster wrote to

Thomas Dawes on July 25, 1809.)[59] The United States declared war on Great Britain in 1812 and called it off in 1815. ("I have examined and collated the radical words in 20 languages, including the seven Asiatic languages or dialects of the Assyrian stocks," Webster wrote to John Jay in June of 1813.)[60]

The *North American Review*, America's first literary magazine, was founded in 1815. ("Wherever the use of *comptroller* exists, it is a reproach to literature," Webster insisted in a December 1816 letter to John Pickering.)[61] James Fenimore Cooper and Washington Irving became America's first significant homegrown novelists, with Irving's *Sketch Book* selling five thousand copies upon its release in 1819. ("I have advanced to the letter *H*," Webster informed Stephen van Rensselaer on November 5, 1821.)[62]

The xenophobia that had animated the national polity at the turn of the century began to subside and transmute into domestic resentments; the cultural and political divide between North and South quickly became a chasm. ("In order to give my work all the completeness to which it is susceptible, I purpose to go to England next summer, if life and health permit, and there finish and publish it," he wrote to Samuel Latham Mitchill in December of 1823.)[63] American culture had evolved on its own without all that much input from Noah Webster.

WHEN the sixty-six-year-old Webster finished his *Dictionary* in January 1825, he was "seized with a trembling," prompted by the relief occasioned by the completion of the work he had feared "he might not then live to finish."[64] So much had changed since he started the project that he must have wondered whether the world even wanted his *Dictionary*. He need not have worried. When the work was eventually published in 1828, it was an international success, with initial editions of twenty-five hundred copies in America and three thousand copies in Great Britain.[65] "I learn however by Newspapers that it meets the approbation of liturary [*sic*] characters," Webster's brother Abraham wrote in a letter from January 1830. "I fondly hope it may be of great use in the learned world & that you may have the consolation of finding that you have not laboured so long to no purpose."[66]

Many of the *Dictionary*'s definitions betrayed Webster's authorial biases. Consider the definition of the word *press*: "A free *press* is a great blessing to a free people; a licentious *press* is a curse to society." Or take the definition of *property*, which Webster used, in part, as a declaration of the morality of copyright and a defense of the authorial-ownership mentality:

> No right or title to a thing can be so perfect as that which is created by a man's own labor and invention. The exclusive right of a man to his literary productions, and the use of them for his own profit, is entire and perfect, as the faculties employed and labor bestowed are entirely and perfectly his own. On what principle, then, can a legislature or a court determine that an author enjoy only a *temporary property* in his own productions?[67]

(This argument is problematic, not least because evidence suggests that Webster may have plagiarized from existing works while compiling his *Dictionary*.)[68]

For the aged Webster, property in literature was more than just an abstract concern. Though the federal copyright law had been a boon to him, its relatively brief term had proven inadequate to his financial needs, forcing Webster to issue new editions of his speller at fourteen-year intervals to maintain an income from its sales. In a letter to the famous Massachusetts politician Daniel Webster—if the two men were related, it was but distantly—Noah expressed his heartfelt wish that Congress would acknowledge that "an author has, by common law or natural justice, the sole and *permanent* right to make profit by his own labor, and that his heirs and assigns shall enjoy the right, unclogged with conditions."[69]

Unlike in the 1780s, when Webster argued for copyright on the grounds that it would encourage the dissemination of books—namely, his own—that would benefit the nation, in later years he was primarily concerned with retaining control over those works for personal profit. Nearing the end of his life, Webster had written many books and wanted to ensure his family continued to benefit from their sale. (He was especially concerned about providing an income for his hapless son, William, whose talents lay in playing the flute and accumulating debts.) The publication of his *Dictionary* made a revision of the nation's copyright laws seem more important than ever—to Webster if no one else. "Few members of Congress feel much interest in such a law, and it was necessary that something extra should occur to awaken their attention to the subject," Webster wrote at the time. That "something extra" would be his own presence on Capitol Hill.[70]

So Webster, revisiting the tactics that had worked so well for him as a young striver, traveled to Washington, DC, in 1830 to lobby for copyright in person. Congress has always been moved by the arguments of famous, well-connected people, and Noah Webster was now both. "I found the members of both houses coming to me and saying, they had learned in my books, they were glad to see me, and ready to do me any kindness in their power," Webster wrote. "They all seemed to think, also, that my great labors deserve some uncommon reward."[71] His works had spoken for themselves—and spoken persuasively.

The bill that ultimately found passage as the Copyright Act of 1831 was an alloyed triumph. The term was still limited, meaning that Congress still declined to see copyright as an inherent, eternal right. In 1814, Great Britain had passed a new copyright law that granted authors a term of either twenty-eight years or the life of the author, whichever was longer; the United States was not ready to go quite that far.[72] Even so, Congress ended up doubling the length of the existing copyright term—from fourteen years to twenty-eight—and retaining the existing fourteen-year renewal period. Moreover, if an author died before the initial term expired, his or her surviving spouse or children could now claim the renewal rights. Meredith L. McGill observed that, by so doing, Congress "redefined copyright as something approximating a personal property right, making copyrights familial and heritable, although only for a single generation."[73]

Webster was content. "My great object is now accomplished," he wrote happily to his wife, and he had good reason for glee: the new law would provide for his family long after his death.[74] But to his son-in-law, William Chauncey Fowler, Webster admitted that his satisfaction was at least partially vainglorious: the reception he received in Washington "convinces me that my fellow citizens consider me as their benefactor and the benefactor of my country."[75] He was, at long last, a respectable gentleman.

WEBSTER'S advocacy on behalf of the 1831 copyright law had once again helped to advance the notion that a book was a product just like a boat, to be bought and sold and protected by law, and that the production of literature should be a profession, rather than just an occasional pastime for men of independent means. But relatively few Americans were racing to ply this new trade.

The average American in 1831, if asked to list ten homegrown authors, would have likely responded with the names of Noah Webster, Washington Irving, and James Fenimore Cooper, followed by a long pause, followed by a quick change of subject. The writer David Leverenz noted that "monetary failure was the norm, for publishers as well as authors, especially for writers of texts not obviously 'moral' or 'useful.' Then as now, most writers tried, failed, and gave up."[76]

American readers generally preferred the works of European

writers to the relatively unsophisticated scribblings of their countrymen. This was partially because Europe's books and authors were
just *better*—or, at least, better known, which to most people has
always been the same thing—and partially because the antebellum
American publishing industry was insufficiently incentivized to nur-
ture and promote domestic talent.

Since American copyright law did not protect those writers whose
works originated outside the United States, publishers could sell foreign books for less than domestic titles. Multiple publishers would
issue multiple editions of the exact same work, with a competitive
advantage accruing to the firm that got its edition printed and distributed first.

The era's book publishers often acted like rival tabloids competing to break news. When a ship carrying British texts arrived in port,
the books were hastened by publishers to printing rooms, where
gangs of compositors worked nonstop so copies could be produced.
Once the book was published, it didn't take long—days, occasionally
hours—before competing American firms took that first American
edition and used it to create their *own* unauthorized editions, which
they sold for lower prices. European books were better, cheaper,
and more abundant than American books. Homegrown authors just
couldn't compete.

It is important to remember, however, that the availability and af-
fordability of these pirate editions engendered a book-reading public
in nineteenth-century America, which would eventually engender a
more mature publishing industry. In this context, what better served
America and Americans? To give authors tight control over their
works, so that they could realize more profit from their good ideas?
Or to build a nation of readers by fostering the flow and dissemination of content? Was copyright a set of social relationships, or was
it a simple property right? Even after passage of the Copyright Act
of 1831, American lawmakers still weren't sure.

Though Webster's letters do not record his thoughts on the

international copyright question, one can speculate on his opinions. Webster had spent his life establishing the parameters of an American-inflected English language that reflected native customs, idioms, and passions. But American literature would never truly reflect the country in which it was created unless authors could profit from their words. By the time of Webster's death in 1843, this line of thought had become increasingly prominent, and the inadequacies of American copyright law had become increasingly clear.

2

A TAX ON KNOWLEDGE

In 1837, the popular nautical-adventure writer Captain Frederick Marryat came to America to do battle with pirates. Marryat had spent twenty-four years in the Royal Navy and had used his experiences as the basis for a series of exciting novels about life at sea. The books were widely read in the English-speaking world, and for them he was relatively well paid; for his 1836 novel, *Mr Midshipman Easy*, Marryat earned about £1,400, worth approximately £144,000 today.[1] It was less than he thought he deserved, and Marryat blamed his publishers for taking "a lion's share" of his sales receipts. "We all have our own ideas of Paradise," he wrote, "and if other authors think like me, the more pleasurable portion of anticipated bliss is that there will be no publishers there."[2]

In Marryat's eschatology, surely a special section of hell was reserved for his American publishers, who could acquire his latest works at no cost and sell the reprinted copies for next to nothing. The American publishing industry was built largely on the backs of British authors like Marryat who were given no succor by federal copyright law. As the American publisher J. Henry Harper put it in his history of his family's pioneering firm, "The Harper brothers saw an enormous reading public in a country of cheap literature, and an immense store of material at their disposition in England, more various and attractive than the home supply, and they resolved to bring the two together."[3]

Nothing could be more simple. And to a foreign author, nothing could be more maddening. American publishers reportedly printed about half a million copies of Sir Walter Scott's Gothic novels, but Scott saw no dividends from this bounty; he died a debtor in 1832, and his peers later argued that "an equitable remuneration [from his American publishers] might have saved his life, and would, at least, have relieved its closing years from the burden of debts and destructive toils."[4] In 1840, Charles Dickens estimated that the combined sales of his works in America had earned him less than £50.[5] "I vow before high heaven that my blood so boils at these enormities that when I speak about them I seem to grow twenty feet high, and to swell out in proportion," Dickens later wrote of the Americans' indifference toward authorial compensation.[6]

Boil and swell as they might, foreign authors had little recourse against American literary piracy, though they sometimes tried to preempt it. Captain Marryat had even gone so far as to provide a Boston publisher with advance copies of *Mr Midshipman Easy* in exchange for royalties on their sales, but this plan was ruined when another, more prominent American publisher immediately issued its own unauthorized edition of the book.[7]

This effrontery galled the captain. His biographer, David Hannay, notes that "the question of money was a very important, sometimes a very pressing one, with Marryat," and that, given the novelist's expensive tastes and speculative imprudence, "it is not overrash or uncharitable to guess that his income was not always adequate to his expenses."[8] Some payment from his American publishers would help ease his financial burden. So Marryat sailed for the United States in the spring of 1837, primarily to tour the country and observe its manners and institutions in the style of Captain Hall, Mrs. Trollope, and other British literary travelers. But he also meant to extract some penance from the American reading public.

He fared poorly. When Marryat went to Washington, DC, to lobby for international copyright, a conversation with a populist Democratic congressman made clear to him that he was unlikely to succeed. "Your authors are very numerous—ours are not," the Locofoco legislator told Marryat.[9] "It is very true, that you can steal our copyrights, as well as we can yours. But if you steal ten, we steal a hundred. Don't you perceive that you ask us to give up the advantage?"[10] This piratical pragmatism was not easily refuted or engaged with, and Marryat left Washington convinced that "we shall never be able to make the public believe that the creations of a man's brain are his own property, or effect any arrangement with foreign countries, so as to secure a copyright to the English author."[11]

The solution, clearly, was for Marryat to disguise himself as an American author. The Copyright Act of 1831 applied only to authors who were "citizens of the United States, or resident therein." But the law didn't define the term *resident*, so Marryat attempted to obtain an American copyright for a new novel, *Snarleyyow*, on the grounds that, as a visitor to the United States, he was currently resident therein. The book was quickly pirated by publishers who simply ignored Marryat's dubious argument. "Capt Marryatt [*sic*] has no more right to this work than we have," wrote one bold Cincinnati publisher. American copyright was not an intrinsic right, but a creation of statute, and if *Snarleyyow* did not qualify for statutory protection, then it was fair game for all. The publisher was not about to be cowed by "a Foreigner attempting to Prevent us from manufacturing Books that are and have been considered and acted upon as common Property."[12]

Marryat gamely tried again, registering two more copyrights on the same basis, this time bolstering his claim to residency by formally declaring his intent to become an American citizen. But the US

Circuit Court in New York questioned the sincerity of the captain's declaration, reasoning that an actual aspiring citizen would not, as Marryat had, so proudly and consistently profess his allegiance to the empire he claimed to be forsaking. (Marryat's true loyalties were most clearly revealed when, at a banquet, he offered a hearty toast to the brave crew of the British ship that had recently sunk an American vessel.)[13] Marryat's copyright application was denied and his claim to residency was rejected.

The rest of Marryat's trip to America was similarly discordant. In his amusing foreword to the 1960 edition of Marryat's *Diary in America*, Jules Zanger observed that before the "tactless and blundering" novelist returned to Britain, he "had been threatened by a lynch mob, had watched his books burned in public bonfires, and on at least two occasions had seen himself hung in effigy by angry crowds."[14] Much as the country's legislators did not want to hear Marryat proclaim their copyright laws unjust, the country's citizens did not care to hear that their cities were ugly, their table manners were crude, and that they spat too much. "What kind of man is Captain Marryatt [*sic*]?" the *Baltimore Chronicle* asked, already knowing the answer. "He is no man at all . . . he is a beast."[15]

OR, more accurately, he was an Englishman, both by nationality and disposition, and that explained much of the disparity. Britain in 1837 was a settled empire with an established class hierarchy, a mature publishing infrastructure, and no expectation that literacy would foster meaningful social mobility. The United States in the mid-Jacksonian era was a rustic nation, and poor. The Panic of 1837, which occurred during the administration of Jackson's successor, Martin Van Buren, precipitated a seven-year recession that "engulfed all classes and all phases of our economic life within its toils," as the historian Reginald McGrane put it.[16] America's cities were largely unvarnished and its vast interior was largely unplumbed; settlers drew sustenance from that which they could hunt, grow, or gather while avoiding wolves, sinkholes, disease, hostile natives, starvation, depression, and other hazards of the frontier. Transit entrepreneurs began to lay America's first railways around 1830, but it took them years to build a reliable network and decades to build a transcontinental line. People traveled via horse, carriage, and inland waterways if they traveled at all.

For the average undercapitalized American, cheap books were the only books, which made the moral case for international copyright something of a hard sell—especially when it was made by citizens of a country that had recently burned America's executive mansion and once tried to tax its tea. If copyright is a social relationship between producers, consumers, and the state, then the relationship that obtained in Britain was far different from its conception in the United States, "where all would be first, and every one considers himself as good as his neighbour," as Marryat wrote in his *Diary in America*.[17] The vulgar national habits that so repelled early nineteenth-century British travelers to America—the bolted meals, the cheap books, the "state of perpetual and disgraceful scuffle" that Marryat balefully noted—were a form of construction debris, generated by a Jacksonian public that had openly renounced the notion that they should toil in quiet obscurity as their betters determined what was best for them.[18]

"How many of the thousands among us who get the last novel of Bulwer, James, or Marryatt [*sic*], for the trifling sum of fifty cents, could make the purchase, if they had to pay one pound eleven shillings and sixpence, or seven dollars, as in London?" wrote the American publisher George Palmer Putnam in 1838.[19] Though that sum was only so trifling because Bulwer, James, and Marryat weren't paid for their work, many Americans argued that their countrymen needed culture more than the British authors needed money. In *Diary in America*, Captain Marryat attempted to characterize American legislators' position on international copyright circa 1837: "It is only by the enlightening and education of the people, that we can expect our institutions to hold together. You ask us to tax ourselves, to check the circulation of cheap literature, so essential to our welfare for the benefit of a few English authors? Are the interests of thirteen millions of people to be sacrificed?"[20] An author's moral right to copyright in his works, populist legislators argued, could and should be superseded if doing so was in the public interest.

Seen from a modern vantage, this official ambivalence toward the sanctity of intellectual property is inconceivable. Today's expansive copyright laws are weighted wholly in favor of content providers and are based on the idea of authors having a natural right to their creations, one that would exist even in the absence of statute. Even though the courts do not formally acknowledge a moral right to copyright in the United States, legislators clearly do: this is the only possible justification for copyright terms that extend far beyond even the longest possible authorial life spans.

But in nineteenth-century America, the concept of intellectual property was not yet sacrosanct—and the interests of readers were not inextricably bound to those of authors. In congressional chambers, lawmakers openly wondered whether international copyright constituted a tax on knowledge and compared literary property to industrial monopoly. American legislators were disinclined to impose this tax on knowledge, to enrich the rentier class at the expense of the laboring public. The nineteenth century was the first Golden Age of Free Content—and denizens of this current digital era would do well to study it, and how and why it came to an end.

CAPTAIN Marryat was not the first literary Briton to try and fail to obtain copyright justice from the American people. A few months before Marryat sailed for the United States, a group of renowned British authors presented a petition to the US Senate enumerating the "deep and extensive injuries" inflicted on them by American publishing customs. The "only firm ground of friendship between nations, is a strict regard for simple justice," the petitioners reminded the Senate, and nothing could be more simple or just than international copyright.[21]

In 1837, these arguments were taken up by Henry Clay, the loquacious and effective Kentucky senator—"Mr. Clay," Marryat noted approvingly, "invariably leads the van in everything which is liberal and gentlemanlike"[22]—who subsequently introduced a draft of an international copyright bill into the Senate on the Britons' behalf. "Of all classes of our fellow-beings, there is none that has a better right than that of authors and inventors, to the kindness, the sympathy, and the protection of the Government,"[23] Clay informed the Senate upon receipt of the British authors' petition. His colleague, the future president James Buchanan, responded by citing a group even *more* worthy of legislative sympathy: "the reading people of the United States."[24] Buchanan encouraged his fellow lawmakers to consider the effect an international copyright law might have "on the acquisition of knowledge in this vast country," and his point proved insuperable. The petition was forgotten, and ultimately, so was the Clay bill.[25]

Buchanan's concern for the welfare of America's bookworms was laudable, but he likely had ulterior motives for making his statement. Philadelphia was home to some of the nation's most prominent "pirate publishers," who uniformly opposed international copyright; as senator from Pennsylvania, Buchanan was obliged to represent his state's interests. In the international copyright debates of the nineteenth century, "the reading people of the United States" often served as a convenient rhetorical shorthand for "the publishers and tradesmen who profit from the production of cheap books." What was said to be best for the former always seemed to correspond with what was inarguably best for the latter.

But just because Buchanan's argument may have been insincere didn't mean it was invalid. Though the absence of international copyright certainly irked British authors, many ordinary Americans considered it a feature, not a bug; a serendipitous legislative lacuna that made it easy to acquire and enjoy the best—or, if they preferred, the worst—of world literature. The reprint copies weren't always of the highest quality, but they were cheap and abundant and generally legible, and they helped create a reading public where once there was none. Since Noah Webster's day, Americans had cherished the idea that they were free to improve their station through hard work and careful study. Education was the poor man's ladder, and for many, cheap books were the rungs.

Periodicals were even more popular than books. The rise of the "penny press" had made newspapers almost universally affordable; the development of scoop-driven reporting and modern interviewing techniques showed the general public that newspapers could captivate. In 1820 there were 512 newspapers in America. By 1840, there were 1,404.[26] These papers took advantage of improvements in printing technology—stereotype printing, which employed reusable metal plates; steam-driven printing presses—to reduce costs and increase production capacity.

But the era's newspapers were rich with partisan vitriol and dubious information. The writer Kembrew McLeod has called the mid-nineteenth century the "Golden Age of Newspaper Hoaxes." Occasionally, to boost circulation, America's penny papers would blatantly invent fantastical stories. In 1835, for example, the *New York Sun* announced the discovery of horned goats and "man-bats" on the moon and rode the ridiculous tale to become the most popular newspaper in America, if not the world.[27] In her recent dissertation, the historian Ann K. Johnson observed that "the very instability of information during this period created a desperate felt need for authority."[28] Works of fiction and history presented substantive and edifying alternatives to the ephemeral insults and scurrilities of the periodical press.

Not only was reading a tool for education and civic engagement in mid-nineteenth-century America, but, as the historian Louise Stevenson observed, it was also "crucial to social bonding. It entertained families and friends while supplying a common vocabulary and world of allusion and imagery."[29] These common cultural allusions were taken from cheap books and newspapers—and they helped build the cultural brain.

"The raven hasn't more joy in eating a stolen piece of meat, than the American has in reading the English book which he gets for nothing," Charles Dickens observed—but one could hardly begrudge the Americans their glee.[30] Copyright was a statutory right in the United States, nothing more, granted by law rather than accumulated custom. A nation that had no natural right to copyright had no moral restrictions against making use of an unprotected British work. Many argued that it was wrong to *not* exploit this bounty. In June 1842, in response to Senator Clay's latest attempt at an international copyright bill, the Philadelphia law-book publishers T. & J. W. Johnson sent Congress a petition noting, "English authorship comes free as the vital air. . . . Shall we tax it, and thus interpose a barrier to the circulation of intellectual and moral light? Shall we build up a dam, to obstruct the flow of the rivers of knowledge?"[31]

The answer from the infringed English authors, inevitably, was an indignant yes. In 1842, Dickens came to America at the behest of an adoring public that had come to love his jolly books, and, by extension, their presumably jolly author. Upon his arrival in Boston, the novelist was welcomed with grand civic festivities, including a skit performed by a prominent comedian in which a "Dickens" character looked benignly on the Americans' copying practices, saying, "But pshaw! I must not quarrel with the 'Trade,' / In golden smiles more richly am I paid."[32]

Dickens's real opinions on the topic could not have been more different, and he spent much of his tour of America chiding his hosts for their unmitigated thievery. In a letter to his friend John Forster, Dickens mocked the Americans' vain presumption that their countrymen's grins were worth more than any banknote. "The man's read in America! The Americans like him! They are glad to see him when he comes here!" Dickens wrote, mimicking the elation and gratitude that American readers apparently expected a British author to feel upon learning that he was popular in the United States. "The Americans read him; the free, enlightened, independent Americans; and what more *would* he have? Here's reward enough for any man."[33] Not only did the Americans routinely cheat British authors out of their rightful earnings, they expected to be thanked for doing so.

British writers took pains to characterize this unhappy custom in stark moral terms, directly equating the unauthorized copying of foreign literature with plunder on the high seas. Like maritime pirates, the authors argued, America's pirate publishers thrived by hijacking others' property for their own personal gain. The piracy metaphor assumed that an author has a natural right to the fruit of his intellectual labor, and that picking that fruit without the author's consent was inherently wrong.

As the 1840s began, these sentiments were routinely echoed by sympathetic American authors, who reiterated their counterparts' arguments while adding some of their own. International copyright was not only the moral choice, the American authors said, but also the patriotic one, as it would improve the quality and quantity of homegrown books. Cheap-book culture vitiated creative ambition and effectively conceded domestic literary dominance to the British. American literature languished—and the lack of international copyright was the reason why.

This, at least, was the refrain of the short-lived, Manhattan-based literary movement known as Young America. In the United States, a prominent Young American named Cornelius Mathews asserted in 1843, "An author is an anomaly; a needless excrescence of nature; a make-trouble and mar-plot, a mere impertinence."[34] In the absence of international copyright, publishers had "set up Notoriety as the great fashionable god, and thrown modest merits far into the shade,"

Mathews grumbled. "A writer must solicit favor and be puffed by the mammoth press . . . his sermon or his lecture must be crammed down the public throat by a newspaper which has its circulation of forty thousand."[35] In order to find any success, Mathews claimed, domestic writers were forced to pander to the tastes of the masses, disgorging derivative stories and low potboilers instead of the more meritorious works that America's aesthetes would have preferred to write and read. The absence of international copyright devalued literature both as a saleable commodity and as an artistic calling. A society that failed to exalt indigenous writers of literary fiction was no sort of society at all.

Today, all that anyone remembers about Mathews was that he was almost universally disliked. In his delightful study of antebellum American literary personalities, *The Raven and the Whale*, Perry Miller wrote that Mathews "was rotund, wore small, steel-rimmed spectacles, bounced when he talked, walked the streets of New York with a strut that nothing could dismay, and delivered himself in an oracular fashion designed to drive all good fellows either to drink or to profanity."[36] A nonpracticing lawyer and writer of bad books, Mathews was utterly convinced of his own genius despite abundant evidence to the contrary, and he spent much of the 1840s singing his own praises in shrill, unmodulated tones. His other great passion was international copyright, a cause that he discredited by virtue of his association with it. Mathews was an outspoken aesthetic nationalist, convinced that international copyright was necessary for the flourishing of American literature, but his ceaseless moralizing on the topic only encouraged his audiences to lose interest in both culture and country.

Mathews must have known that the international copyright omission was but one of many factors limiting America's literary output in the 1840s and 1850s. For one thing, it is hard to grow a literary scene in a nation rent by sectional animus and perpetually verging on economic collapse; for another thing, patchwork distribution networks often limited literary success to the regional level. Yet many fiction writers *did* emerge during this period, if not in the numbers and in the manner that Mathews and his friends would have preferred. They debuted with short stories in magazines such as *Graham's, Godey's, Harper's, Putnam's*. After finding an audience in those outlets, they graduated to novels, which often proved less remunerative than their periodical work. Edgar Allan Poe, Nathaniel Hawthorne, Herman Melville, Harriet Beecher Stowe: they all published their first works during the antebellum era.

These are some of the midcentury American authors best remembered today. Plenty of others have been forgotten. While the naturalistic literature championed by Young America artfully depicted the rigors of the nation's domestic and nautical frontiers, the residents of Strenuous America—the people actually *living* those hard lives—weren't always interested in reliving their hardships in their meager hours of leisure. Sometimes they just wanted escapism. Enterprising writers and publishers could indeed find success in midcentury America, as long as they were willing to cater to public tastes rather than try to dictate them.

Starting around 1860, the Beadle Co. began issuing the first dime novels: nationalistic adventure stories printed in series format on flimsy paper and sold for ten cents a copy. The publisher Frank Leslie built an empire of pulp on a series of illustrated newspapers that lured the public with striking visuals and sensational tales. In the words of publishing historian Madeleine B. Stern, there existed "an all but unlimited demand for entertaining and instructional reading matter in portable format at low prices."[37] And if New York aesthetes were unwilling to provide that reading matter, others would be happy to do so.

Respect the author was the watchword of Young Americans and old Englishmen alike. But what did this motto mean? In a real sense, it meant the desire to tell the American reader what was best for him, to censor the type of material that made its way into his library. "I warn you, I warn you not to withhold this [international copyright] law," Cornelius Mathews admonished his audience in an 1843 address at New York's Society Library. "There are portents already in the sky" of the dark consequences that would inevitably result if international copyright failed to pass Congress.[38] These consequences, to Mathews's mind, were that Americans would continue to write and read bad books—or, expressed less charitably, that men like Mathews would be excluded from influencing popular tastes and defining the canon.

But the very notion of a canon implies the sort of unity and stability that, at the time of Mathews's international-copyright exertions, existed in neither the publishing trade nor the nation at large. As the 1860s approached and America came unmade, the question remained: Was America better served by laws and literature that advanced the cause of the common man, or those that served the interests of the elite? The answer would reveal itself in the decades following the Civil War, as the struggle for international copyright approached its endgame.

IN March of 1879, the American publisher George Haven Putnam took to the pages of *Publishers' Weekly* with news of portents already in the sky, dark visions for the future of the trade. "Within the last year," Putnam wrote, "certain 'libraries' and 'series' have sprung into existence, which present in cheaply-printed pamphlet form some of the best of recent English fiction."[39] These great books were being issued by bad actors: an aggressive new breed of pirate publishers who exploited the continued absence of international copyright to sell unauthorized reprints of British works at unsustainably low prices. These new firms, wrote Putnam, "are not prepared to respect the international arrangements or trade courtesies of the older houses." Their actions spelled disaster for the industry at large.

Few publishers cared as much and as loudly for the health of the industry as did George Haven Putnam, a self-consciously virtuous man in what he considered to be supremely venal times. Putnam was a dutiful son, having abandoned his own ambitions of a career in the natural sciences to work at his father's publishing firm. He was a patriot, a Civil War veteran who had risen from private to brevet major and spent almost five months in Confederate prisons. An enthusiastic joiner, he was active in countless professional organizations and civic causes, as if impelled by honor and breeding to correct the errors of other people's ways. If you read his autobiography, *Memories of a Publisher*, expecting insights into the late-nineteenth-century publishing business, you will be disappointed. If you're looking for ponderous reflections on Putnam's long stint as a member of New York City's grand jury, you will be overjoyed.

Putnam was a standard-bearer for a newly professionalized publishing trade, dominated by family-run East Coast firms with anglicized tastes and aristocratic ambitions. Their leaders were men like Putnam, public moralists and civic volunteers who considered themselves custodians of the commonweal; the sort of men who "led the van," as Captain Marryat might have put it. They believed that the industry had a responsibility to issue *good* books, ones that would elevate their readers, ones that would reflect well on America. Though they were nominally in competition, the heads of these firms all understood that publishing was a pastime for gentlemen, an exercise in cultural virtue.

In an effort to fill the international-copyright void, the major East Coast publishing houses had established an informal system wherein the first firm to arrange to reprint a foreign author's work in America enjoyed exclusive rights to that author's future output. Under this system, the relevant firm could compensate the foreign author without having to worry about being undercut by rival editions. The publishers called this arrangement "courtesy of the trade." Others called it collusion.

By artificially stabilizing book prices, trade courtesy served to dissuade publishers from passing their production savings on to the consumer. Like much else in America at the time, the book business was becoming more efficient. The invention of wood-pulp paper-making, along with a series of national financial panics, had driven the cost of paper to historic lows from 1870 onward. In 1866, fine book paper cost forty cents per pound;[40] by 1871, it cost seventeen cents. Newsprint cost even less. As prices continued to fall over the ensuing years, production capacity simultaneously increased. As of 1867, the fastest American paper mills produced a hundred feet of paper per minute. Over the next thirty years, that rate quadrupled.[41]

As went production, so went dissemination. Despite a series of national financial panics that intermittently halted its expansion, America's railroad network inexorably diminished the nation's frontier, with functional track miles more than tripling between 1850 and 1860.[42] The rise of the railroads not only promoted personal mobility and industrial development, but further diffused print culture. As newsstands opened at railway stations across the country, the American News Company's distribution network kept those stands supplied with disposable, often lurid reading material from the Frank Leslie school of publishing. Writing about a railroad journey she took in 1862, the novelist Louisa May Alcott described how the printed word was hawked in train cars as just another travelers' commodity: "A shrill boy has pervaded the car ever since leaving Portland, shouting, 'Papers, Corn, Books, Water, Lozenges, Sandwiches, Oranges.' "[43]

The white-shoe East Coast publishers were not at first threatened by the rise of railroad literature: Frank Leslie's audiences and George Haven Putnam's were presumed to have little in common other than dimes and sentience. Cheap-book publishers were "more reinforcers of public tastes than creators of them," as Lawrence Parke Murphy put it in Madeleine B. Stern's valuable anthology on the topic; Putnam and his ilk always had higher ambitions.[44] Not until yet another national financial panic in the 1870s prompted many cheap-book publishers to reembrace the republication of highbrow—and uncopyrightable—British literature did the orthodox publishers realize that they had underestimated their populist competitors.

The entrepreneurial Canadian publisher John W. Lovell epitomized this new threat. Lovell was known as Book-a-Day Lovell for the extraordinary fecundity of his printing presses; at peak activity, Lovell printed an astounding seven million books per year.[45] Lovell would identify British books that had already been acquired and popularized—often at significant expense—by the big publishers. Then, once a market had been made, Lovell would reprint the same work and sell it at a low price.

Like the antebellum pirate publishers before them, Lovell and his peers claimed to be performing a public service in disseminating good, inexpensive literature to the unlettered American hinterlands. Their firms openly courted the working-class and impoverished reader. Their books were often shoddily made and astoundingly cheap: published in paperback, circulated by rail and mail. Their loyalties were to themselves and their customers rather than their industry colleagues. And they had no interest in respecting the customs of a club that existed to exclude them.

Though Putnam and his peers cried foul at the cheap-book publishers' disdain for trade courtesy, Lovell scoffed at their self-interested complaints, dismissing the entire conceit as a ploy by established firms to insulate themselves from competition. "I can say to the younger and smaller houses from my own experience, Go in heartily for the 'courtesy of the trade' and—starve," he asserted in a letter to *Publishers' Weekly*. "You will find everything is expected of you and very little given you. As for my part, I prefer to follow the examples that led to success in the past rather than the precepts now advocated to prevent others from attaining it."[46]

The decline of trade courtesy and the concurrent rise of cheap-book dealers such as Lovell corresponded with some broader transformations in American society after the Civil War. Rapid mechanization and industrialization meant that huge fortunes were being made, seemingly overnight, by businessmen exploiting new production technologies. At the same time, rapid urbanization had induced uneducated laborers to emigrate in large numbers to cities such as New York, much to the dismay of many existing residents who thought the city had become overburdened with "the rubbish, the driftwood, and, worst of all, the criminal groups looking for an easy living without labour," as Putnam rather uncharitably characterized these new arrivals.[47] The influx of malleable immigrants and yokels hastened the rise of so-called machine politics—in which votes were delivered to chosen candidates with unfailing reliability—and the political boss, who supervised these effective electoral shenanigans.

To contemporary moralists such as Putnam, the political machines were venal, corrupt entities that empowered the greedy and unqualified while excluding "the honest and intelligent portion of the community."[48] Putnam had a point. Political organizations such as William Tweed's Tammany Hall machine are still synonymous with avarice and municipal corruption because they were, indeed, corrupt and avaricious.

But the obverse of corruption is opportunity. And the political machines and political bosses offered disenfranchised immigrant groups opportunities that might have otherwise been withheld. In *Machine Made*, a recent reevaluation of the Tammany Hall era in New York City, Terry Golway argued that, under Tammany, "the right to vote became, in part, a means to an end rather than an exercise in civic virtue," and that end was access to the sorts of opportunities that the Putnams of the world may have taken for granted—for instance, jobs that weren't demeaning or dangerous, and patrons who would gladly assist their wards instead of comparing them to garbage.[49]

The cheap-book trade resembled Tammany in paperback format. Much like the immigrants who led the Gilded Age political machines, publishers such as Book-a-Day Lovell often hailed from faraway regions and served constituencies with whom the elite were unfamiliar. They approached publishing not as a means of subsidizing and sustaining a specific class of literary workers, but as a means to an end: the expansion of opportunity both for themselves and for the undercapitalized American bibliophile.

In his thorough 1937 University of Illinois master's thesis about the post–Civil War cheap-book trade in America, Raymond Shove briefly profiled one of the most polarizing bargain dealers, an Iowa native named John Berry Alden, dubbed "the Messiah of piracy" by *Publishers' Weekly*.[50] Alden bragged that he offered prices "low beyond comparison with the cheapest books ever before issued." He wasn't exaggerating. In a June 1880 advertisement in *Publishers' Weekly*, he listed his wares at prices twenty to thirty times cheaper than the same titles from traditional publishers: John Bunyan's *Pilgrim's Progress* was available for six cents; Goldsmith's *Vicar of Wakefield* for five cents; and Carlyle's *Life of Robert Burns* for three cents.[51] At prices that low, Americans almost couldn't afford *not* to buy them.

Alden styled his business as a "Literary Revolution," with himself as its leader, seeking to overthrow the cartel that had for too long taxed the common reader. The Literary Revolution was "the most successful revolution of the century," Alden boasted in magazine advertisements, "and, to American readers of books, the most important."[52] Be that as it may, this particular revolution was an unmitigated financial failure. Alden sold his books below cost, which made him at best incompetent and at worst charlatanic. His business was destined to blaze brightly and briefly. But he was more than willing to burn down the rest of the industry with him, and this act of corporate arson helped convince the orthodox publishers that they needed new methods to extinguish his flame.

AS the 1880s began, most Americans harbored little sympathy for the cause of international copyright. "The Congressmen were ignorant of the subject and not easily to be interested in it," wrote Putnam. "To the general public, the idea of property in literature was something of which there was no general understanding and in which there was but a very limited interest."[53]

Many Americans remained desperately poor, and the general sentiment held that "the educational development of the country was more or less dependent upon the possibility of getting the best literature at the smallest cost," Putnam remarked.[54] But Putnam—who inherited his father's business—scoffed at the notion that a nation's "moral and mental development can be furthered by the free exercise of the privilege of appropriating its neighbor's books."[55] The public, he clearly felt, couldn't be trusted to act in its own interest.

"If the teaching of history makes anything evident, it is that, in the transactions of a nation, honesty *pays*, even in the narrowest and most selfish sense of the term, and nothing but honesty can ever pay," wrote Putnam in reference to international copyright.[56] This statement, self-serving and ahistorical, indicates the orthodox publishers' blind, if benign, paternalism, their desire to dictate policies that they claimed were best for others while also, not coincidentally, being best for themselves. "Do I favor an *international copyright*?" the publisher O. J. Victor asked, with rhetorical incredulity, in an April 1879 issue of *Publishers' Weekly*. "Would you imply that any honest man does not? Not to favor it, not to *demand* it, is simply to assent by silence to a wrong equally disreputable and fatal to business probity."[57]

While the cheap-book dealers used revolutionary rhetoric as a sales tactic, the orthodox publishers invoked traditional Christian morality in order to convince the public to support international copyright. "Our idea is to try to bring the matter before the public especially in its moral aspect and to try and educate public opinion in the direction of honesty and fair dealing," Richard Watson Gilder, the editor of the *Century Magazine*, wrote to a friend in 1883 upon the foundation of the American Copyright League, a group comprising many of the era's best-known authors and editors.[58] The league actively recruited literary celebrities—Mark Twain, Louisa May Alcott, William Dean Howells, James Russell Lowell—and then encouraged them to write and speak on the topic incessantly.

Even before the league and its affiliated organizations were formed, George Haven Putnam had been a fount of articles and opinionating on international copyright. Afterward, Putnam's stream became a torrent. As secretary of the American Publishers' Copyright League, a sister organization that was founded in 1887, Putnam diligently penned a minor library's worth of propagandistic speeches, pamphlets, articles, and editorials. ("The trouble with Mr. Putnam," one of his ideological opponents would later say, "is that he writes too much.")[59]

If he wasn't producing his own work, he was promoting someone else's. At the time, many small American newspapers in drowsy towns were accustomed to filling their pages with syndicated material that was shipped from New York or Chicago in preset plates of type.[60] Overworked editors liked this "boilerplate" material because it made their jobs easier. Putnam and his colleagues contrived to insert international copyright propaganda into that boilerplate. As a result, thousands of readers in the American interior were fed a steady diet of international copyright–related content that they would not have otherwise consumed. An added bonus, Putnam recalled, was that legislators from these remote districts "could not get over the impression that an article printed in the home paper must represent, to some extent at least, the opinion of the constituents"; thus, by means of this subterfuge, Putnam won both citizens and lawmakers to his cause.[61]

The league members contacted various ministerial organizations in hopes of convincing their members to work pro-copyright messages into their Sunday sermons.[62] One of the most famous of these homilies, preached by the Presbyterian minister Henry Van Dyke, was later published in book form by Charles Scribner's Sons under the title *The National Sin of Literary Piracy*. "It is altogether idle and irrelevant to talk of 'the lonely rancher in Dakota and the humble freedman in the South,' and their consuming desire to obtain cheap literature," Van Dyke insisted. "The question is, how do they propose to gratify that desire, fairly or feloniously? My neighbor's passionate love of light has nothing to do with his right to carry off my candles."[63]

For all of this literary moralizing, international copyright still almost didn't succeed. Throughout the 1880s, every time legislation was proposed, it was thwarted by manufacturers and tradesmen who feared that the law would be bad for business, and by others who saw copyright as a form of monopoly and expansive copyright terms as a "tax on knowledge."[64] In the first of his two nonsequential terms, President Grover Cleveland encouraged Congress to heed the "just claims of authors" with respect to international copyright; his advice was ignored.[65] "There are some things that make me, at times, ashamed of being an American, and the absence of copyright is one," Gilder wrote to Henry Adams in 1889.[66]

Finally, in 1890, a bill got some traction. Critically, it addressed domestic printers' concerns by stipulating that, to be eligible for international copyright protections, a work would have to be "printed from type set within the limits of the United States." In his report to the full House of Representatives from the Committee on Patents, the House bill's sponsor, one-term Connecticut congressman William E. Simonds, wrote that "the intelligent voice of the whole country asks for the passage of a measure substantially the same as this; authors, publishers, printers, musical composers, colleges, educators, librarians, newspapers, and magazines join in the prayer."[67] At long last, it seemed like that prayer would be soon answered.

Then, disaster: the November elections overturned the House of Representatives' Republican majority, and the incoming Democrats were unmoved by the league's cries for copyright justice. In December 1890, Republican representative Henry Cabot Lodge, a longtime copyright supporter, wrote a blunt letter to Richard Watson Gilder: "There will be absolutely no chance for the [international copyright] bill in the next Congress. Everything depends on passing it now. . . . No time is to be lost." If an international copyright bill was to win the day, it would have to do so before the current legislative session expired and the Democrats took control of the House. March 4, 1891, was the deadline.

The copyright advocates dispatched one of their best men to Washington, DC: Robert Underwood Johnson, secretary of the American Publishers' Copyright League, who, since 1889, had largely abandoned his responsibilities as the associate editor of the *Century* to focus on getting international copyright through Congress. ("But for Underwood Johnson, there would have been no bill," Mark Twain later wrote in his autobiography.[68]) By the time Johnson came to Washington in the winter of 1891, he was already an experienced lobbyist, and he knew what he had to do to bring the copyright bill home.

"I at once organized a systematic appeal to every doubtful Senator, through the newspapers of his State or through constituents or others who we discovered were likely to be influential with him," Johnson wrote in his charming memoir, *Remembered Yesterdays*. "A meticulous study of each man was made from various points of view, and his classmates, clergyman, former business associates and others were enlisted in the good cause."[69] Johnson proved adept at brute-force persuasion, too. After New York senator Frank Hiscock, a Republican, voted to obstruct the bill, Johnson and a colleague representing the typographical unions contrived to have typesetters barrage Hiscock's office with angry telegrams demanding that he reverse his position. The tactic worked: "From that time forward we had no trouble with Senator Hiscock," wrote a satisfied Johnson.[70]

The bill passed the Senate at approximately 1:00 a.m. on

March 4—mere hours before the legislative session would expire. It was sped to the House for final approval, which it received, though not without some agony. Johnson recalled that during the final House debate over the copyright bill, "its noisiest opponent, [Lewis] Payson of Illinois, could be seen by us asleep on a bench at the back of the chamber, his face covered by a newspaper which rose and fell with his stentorian breathing. When at last the roll call came we had an unpleasant quarter of an hour for fear he might awake and rush into the fray."[71] But Payson remained dormant, and the bill passed the House.

Then, disaster, again. Around 2:25 a.m., Samuel Pasco, a Democratic senator from Florida, speaking for a group of senators who believed the bill had been "railroaded through," moved to reconsider the Senate's vote to pass the international copyright bill.[72] (Pasco, oddly, had voted in favor of the bill not two hours earlier.) Democratic senator John W. Daniel of Virginia charged that the measure had been "put through under the instigation and lash of the monopolists who are seeking to aggrandize themselves by it," and insisted that the Senate reconsider its decision and "remedy the wrong which has been done."[73] The league members were briefly struck dumb. "If that bill doesn't pass," remarked Henry Cabot Lodge, speaking for just about everyone in the copyright coalition, "I'll go into a corner and cry."

Finally, at 6:07 a.m., with the legislative day almost finished and Pasco's motion still pending, the spent senators moved to recess briefly and reconvene at 9:00 a.m. to consider the motion and decide the fate of international copyright. "The cup of trembling," wrote Johnson, "was once more at our lips."[74]

By that point, the Senate chamber was nearly empty. The international copyright advocates, bleary and exhausted, had three hours to locate as many friendly senators as possible and herd them back to the chamber to defeat Pasco's motion and pass the bill. Johnson and the publishers William Appleton and Charles Scribner were given the relevant lawmakers' home addresses and dispatched into the rising dawn.

"No one of us will ever forget the experience of that sleepless night," wrote Johnson.[75] "Outside was raging one of the bitterest storms I have ever known. Rain was falling and blowing in gales and freezing as it fell." To make matters worse, the absurdity of the hour and the severity of the weather meant that no carriages were to be found. Despite these adverse circumstances, when the Senate reconvened at 9:00 a.m., the chamber was filled with the friends of international copyright. Pasco's motion failed. With seventy-five minutes remaining in the legislative session, President Benjamin Harrison signed the international copyright bill into law with a quill pen made from an eagle's feather.[76] At once, Johnson cabled Gilder at the *Century* offices in New York: "The bill has been signed by the President! Sound the loud timbrel!"[77]

The legislative process demands compromise, and the final international copyright law was hardly the ideal, unadulterated statute that Putnam and the rest would have preferred. But, still, the law was a meaningful achievement. For the first time ever, mainstream American publishers had unified to lobby for legislation that would protect their businesses and their intellectual property. Their success encouraged them to advocate for even more favorable copyright reforms. In a celebratory April 1891 address to the American Copyright League, the poet Edmund Clarence Stedman put it well when he said, "It cannot be denied that our new Copyright law, if not perfect, wins at least nine-tenths of the battle. If it were quite perfect, perchance we might not feel so sure that this revolution is one of those which never go backward."[78]

THE international copyright law did indeed transform the publishing industry, though not in the way its advocates had expected. Christopher P. Wilson, in his intelligent book *The Labor of Words*, described how, after 1891, American publishing became "a best-seller system which allowed American authors to consistently better their European counterparts for the first time."[79] Trade courtesy crumbled as publishers began to bid competitively for manuscripts and actively market their books to national audiences. Writing for the *Atlantic Monthly* in 1905, the publisher Henry Holt bitterly described a business environment in which a publisher's "interests in authors are narrowed to the moment and to dollars and cents; the dignity and intellectuality possible to his functions—his professional career, as distinct from his money-grubbing career—are destroyed; and his old-time friendships with his authors and 'professional' brethren are reduced to games of dog-eat-dog."[80] American print culture had become, irrevocably, a mass culture.

The international copyright advocates had hoped—at least publicly—that literary professionalization would incubate better books. Instead, literary merit was subsumed in importance to marketability. Wilson observed that the professionalization of authorship in America, "hinging as it did on the legal recognition of [the author's] intellectual property, also coincided with a drive by publishers for greater market predictability and power."[81] Publishers began to act as employers rather than patrons, as businesses instead of benevolent societies. The rhetorical logic of the international copyright debate, through sheer reiteration, led to the development and consecration of the notion of copyright as a property right, and literature as property. And the people who controlled that property resolved to protect it.

3

A COPYRIGHT OF THE FUTURE,
A LIBRARY OF THE FUTURE

In 1895, the people of Boston gathered to christen the nation's nicest public library. Forty-one years earlier, in 1854, the city had opened America's first urban lending library that was free to all and sustained by municipal taxation. Now it had redoubled its commitment to the notion that, as the library's founding trustees put it, "the means of general information should be so diffused that the largest possible number of persons should be induced to read and understand questions going down to the very foundations of social order."[1]

The new Boston Public Library, which cost the city $2,756,384,[2] was a beaux arts palace designed by the architect Charles Follen McKim, with a large interior courtyard, a majestic barrel-vaulted main reading room, and the phrase FREE TO ALL engraved above its main entrance. In an article in the *Forum*, Herbert Putnam, Boston's thirty-three-year-old head librarian, noted that the building represented "a sort of apotheosis of the confidence which the American people have come to feel in the public library as a branch of education."[3]

Though the national supply of cheap British fiction had dwindled after the passage of the international copyright law, American enthusiasm for free or inexpensive culture remained high. The free public library helped meet this continued national interest in frugal autodidacticism. And much as the rise of bargain literature had helped motivate American publishers to unify behind an international copyright bill, the rise of public libraries in America—and all that those libraries represented—would help catalyze the copyright revisions that were to come.

The number of free public libraries in America grew prodigiously in the 1890s and 1900s, precipitated by grants and donations from rich men, such as the industrialist Andrew Carnegie, who hoped to whitewash their fortunes by erecting "a brown-stone buildin' in ivry town in the country with me name over it," as the columnist Finley Peter Dunne wrote in the voice of his Mr. Dooley character. Carnegie began to spend millions of dollars founding libraries for the benefit of the American underclass not a decade after sending armed Pinkerton strikebreakers to quell a labor dispute at one of his steel plants in Homestead, Pennsylvania. (After subduing the Homestead union in 1892, Carnegie subsequently blacklisted many of its members from steel-industry employment.) Six years later, when Carnegie dedicated the library he founded in Homestead, he spoke of his hope that it would serve to "establish a higher code of conduct, a stricter regard for the proprieties of life, and to produce the class of man incapable of anything disgraceful"—such as going on strike, presumably.[4]

The donors' dappled motives notwithstanding, their charity had essentially benevolent results. In town after town across America, public libraries stood as monuments to the idea that the entire nation benefited when information was allowed to circulate freely. In 1895, Herbert Putnam wrote of the American public's conviction that "literature being indispensable, books cannot be too greatly multiplied, or, so far as the readers are concerned, too freely accessible."[5] In an era of rapid mechanical change, the free library qualified as a technological innovation of its own; unrestricted access to books via libraries promised to improve Americans' lives as surely as the automobile or the electric light.

As the nineteenth century yielded to the twentieth, books in America multiplied faster than a child prodigy at math camp. At the Library of Congress, which by law received two copies of every newly copyrighted work, books and other deposits arrived so fast that they were stacked on the floor in great hoarder's piles. "By 1897," wrote Library of Congress historian John Y. Cole, "the collections of the Library had overflowed into over a dozen different locations throughout the Capitol, including the attic and the cellar."[6]

It was an era of cultural abundance. The periodical press continued to trump literature in popularity and circulation. America boasted approximately 2,226 daily newspapers and magazines in 1899, up from 254 in 1850,[7] and publishers profited from new printing improvements that raised production capacity while lowering production costs.[8] The boundaries of the known cultural universe expanded to include new commercial and artistic technologies. Some, such as motion pictures and recorded sound, composed brand-new artistic media. Some were used to change extant art forms—not only did the electric light transform live performance, it also made it easier for homebodies to read themselves to sleep. Some, such as the telephone and the mimeograph, revolutionized the transmission and dissemination of information. All of these new technologies emerged in the decades preceding 1900, and with them emerged a new cultural economy.

"Managing the production, marketing, and distribution of goods in this new national arena required a new cadre of administrators and professionals," the cultural scholars Carl F. Kaestle and Janice A. Radway have observed. "In creating and training such a cadre, corporate capitalists created a virtually new social class, one that Marx had not anticipated."[9] These administrators' livelihoods were supported by proprietary media, yet they were rarely creators themselves. Their rise to prominence was inevitable. As Christopher P. Wilson put it, this was an era when literature "could be conceived as a product of labor rather than romantic inspiration."[10] The product of labor is property. Information wants to be expensive.

But the rise of the public library in America had institutionalized the notion that information wants to be free—and that there is national benefit in its liberation. Libraries "represent the accumulated experience of mankind brought to our service," Putnam wrote in the *North American Review* in 1898. "They touch the community as a whole as perhaps does no other single organized agency for good."[11] While librarians imagined new ways for preserving and sharing this new cultural abundance, this emerging cadre of middlemen were determined to find new ways to make it pay.

ON December 5, 1905, as part of his lengthy annual message to Congress, President Theodore Roosevelt advocated a total revision of the nation's "confused and inconsistent" copyright laws to better meet the "modern conditions" of the machine age. The man charged with shepherding this law was Herbert Putnam, the former head of the Boston Public Library, who had served as librarian of Congress since 1899. Putnam was a lawyer by training and a bibliophile by birthright. He was one of *those* Putnams, a scion of the New York publishing family that, for decades, had promoted literature as a sort of civic vitamin, and life as a civic duty. As a child, his elders described him as "plain but interesting," and as a young man, he initially seemed disinclined to reach for anything more.[12]

"I have not indulged an appetite for exploration," he informed his far-flung former Harvard classmates in 1886, when he was twenty-four. "No books, pamphlets, articles, addresses, or other literary progeny rest upon my conscience."[13] Instead, he chose to spend his life managing other people's output, intent on improving the world from behind the circulation desk. In Boston, wrote historian Jane Aikin, Putnam extended the library's hours of operation and opened one of the nation's first dedicated reading rooms for children. He appointed women to prominent administrative positions and implemented performance-review policies that "favored the enterprising and accomplished over the less educated and less energetic."[14] Putnam was small and gouty and meticulous; his brother John later observed that "the learned gentleman," as he called him, "fully believes it is his duty to endeavor to improve on existing methods if these seem, in any particular, imperfect."[15] He was a natural choice to lead the nation's largest library into the twentieth century.

"A Conservative and Cautious Man," read the *Washington Evening Star*'s appraisal of Putnam on the day of his arrival at the Library of Congress.[16] "He loves hard work, he says, and has come to Washington with full appreciation of the fact that there is plenty of hard work before him." In the capital, he sought to modernize the Library of Congress and expand its collections—literary and otherwise. "A book is not the only nor necessarily the most effective vehicle for conveying knowledge," Putnam wrote in the *North American Review* a year before he came to Washington. "And photographs and process reproductions are now part of the equipment of a public library almost as conventional as books."[17] Photographs, lithographs, sound recordings, piano rolls, the telharmonic dynaphone of Dr. Cahill—all were elements of culture, and all fell within the ambit of the modern library. In 1904, the Library of Congress acquired its first phonograph recording, a cylinder featuring the voice of Kaiser Wilhelm II.[18] Many more would follow.

As new forms of content emerged, legal questions abounded: How far should copyright extend? Whose interests should the laws protect—consumers or producers? And what sort of producers? Should copyright be used to encourage new industries, or to protect old ones? To ask these questions is inevitably to find them tedious and complicated—which explains why copyright statutes only get revised once or twice a century.

"If you can get the interests represented, the interests practically concerned, to agree upon a measure, then we will endeavor to deal with it," the Senate Committee on Patents airily told Putnam, who supervised the office of the register of copyrights. "We do not want to have a series of hearings in which adverse interests shall appear and oppose and demand, and so on."[19]

Thus tasked with devising a new omnibus copyright bill that would be palatable to congressional and business interests alike, Putnam decided to bring the relevant parties together and convene a series of copyright conferences at the City Club in Manhattan— the old headquarters of the men who had fought for international copyright. But circumstances had changed since those debates, and copyright advocacy was no longer the sole purview of authors, editors, publishers, and other genteel literary entrepreneurs.

When Putnam called roll at the beginning of these conferences, he was answered by men representing myriad creative professions: lithographers, photographers, theater managers, advertising men, telephone-directory publishers. Not all of these representatives were themselves active practitioners of the arts. The public ethicists and civic activists who had masterminded the international copyright campaign shared few personality traits with these newer arrivals, men such as the headstrong George W. Furniss of the Music Publishers' Association, who, early on, admitted that "it was unusual for us to be in the presence of eminent people," and soon demonstrated the truth of that assertion with his plainspoken contributions to the convocation.[20]

Putnam, whose professional life had been dedicated to collecting and distributing books for the public benefit, made for an odd collaborator with men such as Furniss, who was most concerned with securing his constituents the copyrights to their tinny, middlebrow nonclassics, and putting sheet-music pirates in prison. "We have just now succeeded in locking up a man in your prison here," Furniss announced at one point. "He will probably get a term of one year. When he gets out, we will probably get him in New Jersey and lock him up for ten years."[21]

The sessions were characterized by this sort of plain and aggressive talk. The moral overtones of the earlier copyright campaigns had been muted; at the beginning of the first conference session on May 31, 1905, Putnam noted, "In our proposals for copyright we had, I think, better leave the moral yearnings to a later generation or at least a later session of Congress."[22] Instead, the attendees worked to entrench the idea of intellectual property as, indisputably, property. Plenty of disputes erupted along the way, including one long and odd digression about whether sculptors and painters seeking copyright protections should have to inscribe a copyright notice on "some visible portion" of their works (the answer, ultimately, was yes), but all participants were basically in accord on several essential points: copyright terms should be long and easily renewable; scofflaws should face civil or criminal penalties; the public benefit in copyright was identical to the author's benefit; and those who felt differently were probably pirates.

Once the conferences had concluded, Putnam and the staff of the Copyright Office devised some draft legislation and prepared it for Congress. In June and December of 1906, the joint House and Senate Committees on Patents held hearings on the bill, and its framers convinced various cultural celebrities to speak on its behalf. Samuel Clemens, better known by his pseudonym, Mark Twain, delivered a rambling, vaguely funny speech on the merits of extended copyright terms: "Make the limit the author's life and fifty years after, and, as I say, fifty years from now they will see that that has not convulsed the world at all. It has not destroyed any San Francisco. No earthquakes concealed in it anywhere. It has changed nobody. It has merely fed some starving author's children."[23]

The composer and bandleader John Philip Sousa claimed that the sale of recorded music—from which composers received no royalties—imperiled not just his own livelihood but the entire American musical tradition. Sousa cut perhaps the most memorable figure at the hearings, with his charming testimony and comical insistence that, in the absence of new copyright laws, human vocal cords would soon atrophy, then ultimately vanish, like early man's vestigial tail.[24] "These talking machines are going to ruin the artistic development of music in this country," Sousa warned. "When I was a boy—I was born in this town—in front of every house in the summer evenings you would find young people together singing the songs of the day or the old songs. To-day you hear these infernal machines going night and day. [Laughter.] We will not have a vocal cord left. [Laughter.]"[25]

But Sousa's imaginative evolutionary theories failed to close off debate on the relative merits and demerits of the "infernal machines." George W. Pound, who represented two musical-machine concerns, one in Cleveland, Ohio, and the other in Tonawanda, New York, claimed that "as a matter of fact every composer in the land and every music-publishing house in the land is glad to get the advertisement following from the mechanical reproduction of their music. It is regarded in the trade as the best assistant to the sales of their music of any form of advertising."[26]

Others noted that some of the same composers who testified that new copyright protections were needed in order to prevent unauthorized recordings of their works were simultaneously corresponding with phonograph companies, requesting that they produce recordings of their latest sheet-music publications.[27] "Mr. Sousa himself does not scorn, as he pretended to the other day, these 'infernal talking machines,'" charged S. T. Cameron of the American Graphophone Company:

> He to-day is under contract, and he plays into these "infernal machines" with his band, and he is contributing, as he told you a few

days ago, to stifle these "beautiful young voices that now have disappeared throughout our city and our land." [Laughter.] He does it for the almighty dollar. That is what he is after, and he frankly told you so.

Mr. SOUSA. I am honest, anyway. [Laughter.][28]

Inventors and entrepreneurs from across America traveled to Washington to protest the bill, which they considered a conspiracy devised by moneyed interests to enrich themselves at the expense of emerging competition. This was a common theme in the era's polity. The copyright deliberations came at the height of the trust-busting years, when President Roosevelt was swinging his big stick at industrialists' knees, and the American public imagined tentacular corporate machinations behind every closed door. In this environment, a new copyright law was bound to be seen as a monopolist's ploy. "I say that we can hope for neither glory nor popularity with this measure," Putnam predicted in December 1906. "Out of favor with the proponents, denounced by the opponents, unpopular with the public, and probably more or less discredited by the committee—that was the prospect!"[29]

Putnam was prescient. Independent operators railed against the "star-chamber proceedings" in which the bill was devised and saw malfeasance in the apparent exclusion of outsiders from the proceedings. The most-cited monopolistic copyright conspirator was the Æolian Company, a player-piano manufacturer based in New York, whose presumed intentions seemed as sinister as anything Standard Oil had ever devised. The Æolian Company had convinced various music publishers to assign it the exclusive right to reproduce and distribute their music in piano-roll format if piano rolls were found to be copyrightable content. Those piano rolls would, of course, only be compatible with an Æolian pianola, and this exclusionism would drive all other manufacturers out of business and inhibit innovation in the player-piano industry.

The litany of complaints against the professional practices of the Æolian Company and the presumed conspirings of the copyright advocates stretched on, ultimately becoming the early focus of the Senate copyright hearings. "If the inventors of this country knew what was in this bill, there would be enough here to fill up every room in this great building, but they do not know it," exclaimed the inventor G. Howlett Davis.[30] If the copyright bill went through unamended, "the Æolian Company and the concerns affiliated with it will have millions of dollars turned into their coffers. And the net result is that the public will pay," insisted the attorney John J. O'Connell.[31] The copyright advocates "profess great interest in the public and in the cause of musical education," Pound said, "but it seems to me that the public is the one and the sole element which has been omitted in their consideration."[32]

Though Putnam expected neither glory nor popularity for his copyright labors, he certainly did not expect to have his sincerity called into question. "What was the *consideration*, the inducement to the Copyright Office to enter into this conspiracy?"[33] the learned gentleman asked in frustration after a particularly pernicious attack on his probity. There was no conspiracy—or, at least, none of the tinfoil-hat variety. The law's framers' narrow conception of what constituted the public's best interest was less a function of malice or sedition than of selection bias. The bill had been conceived and written by "the most representative organizations that we could think of or that were brought to our attention as having practical concern in the amelioration of the law, but especially, of course, those concerned in an affirmative way—that is to say, in the protection of the right," Putnam acknowledged at the outset of the hearings.[34] One man's right is another's wrong.

The committee's hopes of avoiding a long and disputatious copyright debate had been in vain. The bill, first brought to Congress in 1906, would not be signed into law until 1909. The new Copyright Act increased the standard copyright term for new works to twenty-eight years, with a twenty-eight-year renewal period, and expanded the universe of copyrightable content. Most notably, as a compromise between musicians and recorded-music concerns, the act created the "compulsory license," by which record companies and mechanical-music manufacturers would pay musical composers a small royalty for each song of theirs that had been recorded and sold.

The law remained more or less unchanged for nearly seventy years. "We want a copyright of the future," Sousa said, one that did not differentiate between a written work and a musical recording in terms of the protections offered to its authors.[35] He and his fellow musicians got one, eventually. But the vexing thing about the future is that it soon becomes the past.

HERBERT Putnam served as librarian of Congress for forty years, retiring in 1939 but remaining in Washington as librarian emeritus. In a 1903 commencement speech to the graduates of Columbian College in Washington, DC, Putnam had proclaimed that "the world is a cheerful world today, and the most interesting world that ever was."[36] As the years passed, the world grew more interesting and less cheerful. During World War I, Putnam headed the American Library Association's Library War Service, supplying soldiers with books on relevant subjects such as "French history, mechanics, topography and strategy in war, self propelled vehicles, hand grenades," and seemingly random topics such as stage directing and American business law.[37] "The Library has its part to play [in the war effort]—an indispensable one. Its efficiency *must* be maintained," he told his staff.[38]

For decades, Putnam presided over a regular "Round Table" luncheon society in the Library of Congress's dining room, hosting guests from the worlds of business, politics, academia, and the arts. There he watched as silent films became talking pictures, produced by great film studios; as piano rolls and wax cylinder recordings gave way to vinyl records and radio broadcasts. Motorized public trolleys came to Washington, and Putnam became a dedicated customer. When he didn't ride to work, he walked; on at least one occasion, he reportedly roller-skated from his house on O Street NW to his office at the library.[39] He acquired for the Library the personal library of the stage magician Harry Houdini; he opened a center for Hispanic studies. In 1928, the *New York Times* called him "not only a 'model librarian' of a library which he has made a model. He is an outstanding citizen of the world."[40]

In 1939, on Putnam's fortieth anniversary as librarian of Congress, President Franklin D. Roosevelt wrote him a letter announcing his "unshaken conviction that democracy can never be undermined if we maintain our library resources and a national intelligence capable of utilizing them."[41] Putnam retired six months later, two years before the United States would enter and be forever transformed by its participation in World War II. In 1954, he told the *Washington Post and Times Herald* that "I keep alive intellectually by contact with my friends at the club. Men of science are the nucleus. They are men whose language I don't speak. But to listen is an education."[42]

Putnam died in 1955 in Woods Hole, Massachusetts, at ninety-three years of age. The world he left behind was caught in social and technological transition, just as it had been when he first arrived in Washington. The Soviet launch of the satellite *Sputnik* in 1957 precipitated a thirty-year game of Keeping Up with the Khrushchevs during which the US government directed unprecedented sums toward laboratories, universities, and research agencies charged with devising technologies that would help America overtake its Russian rivals. In 1961, America launched the astronaut Alan Shepard into outer space—one month *after* the Russian cosmonaut Yuri Gagarin made his own star tracks.

If the United States at the time couldn't quite match the USSR in interstellar accomplishments, it could at least demonstrate its superiority on the ground. On March 13, 1962, President John F. Kennedy addressed the House of Representatives with festive news. The 1964 New York World's Fair was approaching, and Kennedy had endorsed the construction of a federal pavilion that would celebrate everything America had to offer the world of the future. The theme of the pavilion was Challenge to Greatness, and in his speech to Congress, the president explained that the proposed pavilion would "present to the world not a boastful picture of our unparalleled progress, but a picture of democracy—its opportunities, its problems, its inspirations, and its freedoms."

By the time the fair opened in 1964, event organizers and sponsors had transformed Flushing Meadow Park into a showcase for American industrial progress, an idealized, chrome-plated version of a future that would soon be on sale at a department store near you.[43] The fair was dominated by American corporate pavilions, most of which attempted to embody the fair's subtheme: Man's Achievement on a Shrinking Globe in an Expanding Universe. The Du Pont chemicals pavilion offered forty-eight performances per day of the "Wonderful World of Chemistry" musical revue, which was followed immediately by "about two dozen demonstrations of startling uses for products made by Du Pont."[44] The Formica World's Fair House promised that housekeeping would be a dream just as soon as all the walls in your home were covered in plastic laminate.[45]

Like every other pavilion at Flushing Meadow Park, Challenge to Greatness was a sales pitch—but instead of touting household chemicals or synthetic walls, the pavilion advertised the American way of life. Visitors to Challenge to Greatness were clobbered by appropriate symbolism from the moment they arrived. Upon approach, they were greeted by the likeness of a gigantic eagle, which presided imperiously over the steps leading up to the pavilion. Next, they viewed a short film titled *Voyage to America*—an ode to the immigrant spirit. The centerpiece was an educational theme-park ride called the American Journey, in which the star of the television program *Wagon Train* intoned a sentimental script by the novelist Ray Bradbury that encouraged visitors to "move with heart and purpose toward a tomorrow of your own choosing."[46]

The idea that Americans are free to choose their own tomorrows underpinned the Challenge to Greatness. The national emphasis on bootstrapping, on self-improvement, was just as strong in 1964 as it had been in Webster's time, and intellectual mobility remained as important to America's success as economic and physical mobility. So it was perhaps appropriate that the tour of the American pavilion, which began with a bald eagle, concluded with another symbol of the nation's untethered spirit: a library.

In 1895, Herbert Putnam wrote that the rise of the public library in America had advanced "a novel idea: that a book has an active as well as a passive duty to perform; that it should not merely be hospitable to those who come to seek, but should itself go forth, should seek out the individual and impress its stored-up activities upon him."[47] Ever since then, progressive librarians have worked toward the establishment of a library where the knowledge you needed could find *you*—or, at the very least, would always be at your disposal, regardless of race, gender, creed, or bank balance. As Kennedy himself had put it, in a quote prominently displayed in the Challenge to Greatness pavilion, "Books and libraries and the will to use them are among the most important tools our nation has to diffuse knowledge and to develop our powers of creative wisdom." Part of the challenge would come in trying to use new technologies to meet this ideal.

Library/USA, the final stop on the visitor's tour of the extravaganza, was a large exhibit sponsored by the American Library Association. "We anticipate that [the visitor's] curiosity will have been aroused at many points" in his trip to the fair, the organizers wrote, and the Library/USA exhibit was there to sate that curiosity. It featured a team of reference librarians ready to answer visitors' questions, a full working children's library called Children's World ("the man who reads begins as a child who reads"), an inside look at the president's library, and many other bibliocentric displays. Perhaps most exotic of all, the exhibit featured a UNIVAC 490 Real-Time computer that could, within seconds of being asked, provide visitors with detailed information about many of the topics broached in the preceding exhibits.

The UNIVAC epitomized one of the fair's general themes: Peace Through Understanding. As the wall text of the Library/USA exhibit noted, "In the last 20 years mankind has acquired more scientific information than in all previous history." But the resources were not equally distributed. "Good libraries are not available to all," the curators observed. "More people will need more knowledge." Mechanized libraries, they submitted, would facilitate the frictionless sharing of information, which would in turn help close the international knowledge gap.

Upon leaving Library/USA, visitors were presented with a UNIVAC printout titled "Challenge to Greatness," which was supposed to be "symbolic of the continuing challenge to increase each person's free access to sources of information." One can hardly think of a better closing image with which to leave pavilion-goers—a better summation of the installation's themes, of one way the country might meet the challenges of tomorrow—than this peek at what could only be considered the library of the future.

One year after the World's Fair began, in August 1965, the Massachusetts Institute of Technology collected some of its brightest minds in Woods Hole, Massachusetts, to help make the library of the future a thing of the present. Microfilm and digital computers promised to revolutionize information storage, retrieval, and transfer, and MIT wanted to integrate these technologies into the infrastructure of the modern library. For the next five weeks, the MIT scholars, alongside various visitors and special guests, discussed plans for a prototype digitized library system far more advanced than anything found in Library/USA. They called their undertaking Project Intrex.

The name Project Intrex—*Intrex* was short for "information transfer complex"—sounds like something out of a Cold War spy movie, and that's not too far from the truth. The Independence Foundation, a group that itself drew funding from the CIA, sponsored the Woods Hole conference. The project was headquartered at MIT, which had already effectively become a research arm of the federal government. The project's top men had spent years working at government-sponsored research agencies, such as Lincoln Lab, the Advanced Research Projects Agency (ARPA), and the RAND Corporation. An implicit objective was to employ the technology that Intrex developed to help the United States thwart the Soviet menace; from this perspective, libraries were important insofar as they could help scientists build new and better weapons more quickly.

The keynote speaker who launched the Intrex conference that August was, appropriately, the godfather of what today we call Big Science. Vannevar Bush was no stranger to government-university partnerships. First as Franklin D. Roosevelt's science adviser during World War II, and then as the motive behind the creation of the National Science Foundation, Bush, as much as anyone, was responsible for the militarization of American academic science. In 1965, he was seventy-five years old, and his long and complicated career was nearing its end. His brief remarks to the group at Woods Hole were wistful: "I merely wish I were young enough to participate with you in the fascinating intricacies you will encounter and bring under your control."[48]

Vannevar rhymes with *believer*, and when it came to government funding of scientific research, Bush certainly was. He was also a lifelong believer in libraries, and the benefits to be derived from their automation. In 1945, he published an article in the *Atlantic Monthly* that proposed a rudimentary mechanized library called Memex, a linked-information retrieval system. Memex was a desk-size machine that was equal parts stenographer, filing cabinet, and reference librarian: "a device in which an individual stores his books, records, and communications, and which is mechanized so that it may be consulted with exceeding speed and flexibility."[49] The goal was to build a machine that could capture a user's thought patterns, compile and organize his reading material and correspondence, and record the resulting "associative trails" between them all, such that the user could trace his end insights back to conception. Imagine the thrill, for example, of plumbing the neural archives of the man who invented whistling, and being able to trace the development of that concept, step by step, all the way back to that magic moment when he realized that he might be able to improve on the hum.

The world of 1945 wasn't quite ready for Memex, or quite capable of building it. But the world had changed in the intervening decades, and in 1965 it looked as if the Memex concept might finally come to life. In his introductory remarks to the Intrex conference, Bush maintained that computerized libraries would transform society, and that Project Intrex would "influence, perhaps revolutionize, the methods of every professional group—in law, medicine, the humanities. It will support every phase of our general culture. I believe very few scholars today realize what this could mean. I am sure the general public does not realize, for instance, that success in this program could mean as much to their well-being, their health, as has been produced by the power of antibiotics."[50] Even the most pragmatic men couldn't help but sound idealistic when it came to libraries and their potential to change the world.

AS librarians and scientists were touting technology's ability to ease and improve the spread of information, a group of copyright lawyers were meeting in Washington, DC, to discuss ways to restrict that data. The federal copyright statute had gone largely unrevised since 1909 and was far behind the times. "I am confronted daily with what are now being called the 'information explosion' and the 'communications explosion,'" said L. Quincy Mumford, librarian of Congress, in 1965. "It is obvious to me that these revolutionary developments carry with them a profound challenge to creative endeavor, and that our antiquated copyright law must be revised to meet this challenge. The longer this task is delayed, the harder it will be to accomplish, and the more serious will be the loss for future generations."[51]

Mumford asked the register of copyrights, a genial man named Abraham Kaminstein, to convene a panel of eminent copyright lawyers to draft yet another copyright law of the future. They were, of course, determined to avoid the mistakes of the past. The panel's goal was both to extend copyright terms and to expand copyright protections to all sorts of new media. Antediluvian entertainments, such as silent films and piano rolls, had given way to record players, jukeboxes, 3-D movies, and other new technologies, all of which had prompted innumerable copyright questions. Did artificial flowers deserve copyright protection? What about "the smellies"—the short-lived breed of films that blasted relevant odors into the theater? And what about machine-to-machine communication? Or even extrasensory perception?

"This is serious," said Herman Finkelstein of the American Society of Composers, Authors and Publishers (ASCAP). "It's serious because, when Scrutton wrote his book on copyright back in the last century, he said, 'Beware. Those people who talk about a short term of copyright may be leading us into communism.'"[52] This cultural-siege mentality, as much as anything, seems to have motivated the copyright advocates as they set about drafting the new law. Irwin Karp, representing the Authors League of America, spoke for many when he observed, "We're writing a law now that may last another fifty years. Even though all of the evils and dangers that we raise may not exist in their fullest extent today, they may be existing five years from now."[53]

The challenge for the panel, then, was to devise a bill specific enough to protect the rights of existing stakeholders, but broad enough to cover technologies that did not yet exist. One of the most vocal participants, an attorney named John Schulman, set the tone in the early going when he expounded upon a copyright truth that seemed self-evident: "Why would a man write a book—a textbook or something else—unless he had an opportunity of improving himself economically? Why do we hesitate to say that we are talking about the right of an author to earn a livelihood? That's what we're talking about."[54]

And that was the prevailing sentiment expressed by the new bill's supporters, both in private panel discussions and, later, in congressional testimony. Copyright was depicted as a paragon of the American way of life—which, like so much else about America at the time, seemed under attack from chaos agents willfully rejecting the nation's fundamental values. In May 1965, Will Dillon, the eighty-seven-year-old co-composer of the sentimental barbershop standard "I Want a Girl (Just Like the Girl That Married Dear Old Dad)," urged Congress to extend copyright terms. In his brief testimony, he pulled on every imaginable heartstring, citing Winston Churchill, Arlington National Cemetery, World War II, even trotting out his granddaughter, Georgia Ann Murdock, to read a prepared statement, perhaps reasoning that everything sounds more sincere if said by a granddaughter.[55] Few of her grandfather's songs had "survived the ravages of time and changes in public taste," Murdock read. If the copyright on "Dear Old Dad" was allowed to expire, Dillon would lose the mainstay of his income. Allowing old songs and books to fall into the public domain served only to impoverish America's elderly songwriters and their adorable progeny.

Not everyone was convinced by these arguments. "We keep hearing about the rights of various groups of authors, producers, broadcasters, and so on," said Ralph S. Brown Jr., a Yale Law professor, and one of the few voices of moderate dissent among those debating and shaping the bill. "Occasionally an attempt is made to define 'the public.' A few moments ago a definition of 'the public' occurred to me—that 'the public,' from the standpoint of these discussions, is anybody from whom another nickel can be milked for copyright owners."[56]

Brown was decidedly a minority voice. In a remarkable sequence early on, one participant after another challenged the utility, public benefit, and very legitimacy of the public domain. "I don't believe that, just because a work goes into the public domain, that the public is benefited," declared John R. Peterson of the American Bar Association. "Stated in our Constitution itself is the fundamental public interest there, to stimulate creativity. I don't think it stimulates creativity to tell people that they are welcome to copy what someone else has written."[57] When a work entered the public domain, according to this logic, it represented a loss of valuable rights for the author rather than a gain of valuable rights by the public.

If neither the public nor creators benefitted from limited copyright terms, then who *did* benefit? The answer, as always, was pirates. Pirates have no place in polite society: they wench, they carouse, they make their victims walk the plank. Worst of all, pirates use violence to help themselves to other people's property. The term *pirate* groups copyright infringers in the same category as Blackbeard, Captains Kidd and Hook, and other notorious maritime villains. It's hard to recover from that initial mental association, from the image of rogue bands of printers, tape dubbers, and photoduplicatrixes roaming the oceans of copyright, ruthlessly plundering innocents for their personal gain.

In the prelude to the Copyright Act of 1976, pirates proliferated. They were in the classroom, masquerading as harmless teachers, their most enduring lesson being a profound disrespect for private property. They loitered in lending libraries; they seized the control rooms of local educational-television stations; they lurked in basements with tape decks, illegally copying copyrighted music and distributing it without paying the composer for the privilege. They were especially to be found in the corner taverns, dancing their mirthless pirate jigs to songs blaring from the jukebox.

To resolve the dispute between composers and the infernal-talking-machine proprietors, the 1909 framers created a public-performance clause, wherein composers were compensated if their works were exhibited in public. But the framers decided that a song played on a jukebox did not constitute a public performance; in 1909, jukeboxes were more like primitive personal stereo systems.

Since that time, however, jukebox technology had improved, and by 1960, America's taverns and hamburger joints had approximately half a million jukeboxes, each earning a small fortune in small change per year, not a cent of which went to the composers of the songs it played.[58] The jukebox owners had no reason to retain their exemption from the public-performance clause, copyright holders argued. The shortsightedness of legislators in 1909 in not anticipating the rise of Wurlitzer Nation didn't mean that Elvis and the Beach Boys should miss out on royalties, the panel felt. They wanted the loophole closed.

The photocopier, like the jukebox, inspired fear and loathing among members of the copyright committee. Though mechanized document-duplication devices had existed since the 1920s, the invention in 1959 of the one-piece photocopier, or Xerox machine, made photoduplication a veritable pleasure rather than a rage-inducing chore. Now, with minimal effort, anyone could duplicate any bit of printed content he or she desired. Much of the discussion during the meetings was devoted to strategies for thwarting the rise of the Xerox machines.

The arguments that were made decades before about piano rolls recurred in this new context, with some lobbyists for educational organizations claiming that authors and publications benefited from having their articles photocopied and distributed, insofar as such activity might actually persuade the duplicators to subscribe to the journals in question. The *Chicago Tribune* opined, "The school mimeograph should be viewed not as a piratical rival to the trade publisher but as a helpful unpaid publicity agent who helps publishers' long-term sales."[59]

That argument didn't take root. The rise of photoduplication technologies that facilitated the rapid spread of information merely underscored the fragility of copyright holders' claims that intellectual property was indistinguishable from regular physical property. "We know that volumes of information can be stored on microfilm and magnetic tape. We keep hearing about information-retrieval networks," former senator Kenneth B. Keating told Congress in 1965. "The inexorable question arises—what will happen in the long run if authors' income is cut down and down by increasing free uses by photocopy and information storage and retrieval? Will the authors continue writing? Will the publishers continue publishing if their markets are diluted, eroded, and eventually, the profit motive and incentive completely destroyed? To pose this question is to answer it."[60]

IN her interesting book, *Who Owns Academic Work?*, Corynne McSherry made a distinction between a fact and an artifact.[61] A fact is one of "nature's creations," an ownerless piece of knowledge or data that belongs to the public. The periodic table, for example, is a fact: the tabular relationship between the chemical elements belongs to no one person, but instead is common knowledge on which all are free to draw and build. An artifact is a proprietary derivative of a fact: a specific and unique expression of fact or fancy. A colorful, uniquely designed poster of the periodic table, then, is an artifact, and the poster's designer owns the right to sell copies of that poster and prevent you from reprinting it without permission. Facts are in the ether and cannot be deaccessioned—they belong to everyone. Artifacts are most often found in museums, cordoned behind velvet ropes, with direct access restricted to authorized personnel.

Another useful concept when attempting to understand the various points of view on copyright and intellectual property issues is the difference between gift economies and market economies. In a gift economy, a person gains influence and stature by the skill with which he gives and receives gifts; the gift, which is freely given with no expectation of recompense or tangible reward, is the primary medium of exchange.

The university has traditionally been a gift economy. Research is produced there and disseminated to the world; the academic gift-giver benefits when his theories and discoveries gain acceptance and adoption. Others then use his work to make new discoveries, from which the initial gift-giver in turn derives benefit. Nonacademic work, however, exists in a market economy, where cultural artifacts are bought and sold, and the creator benefits by reaping financial reward from the sale of his work. The creator is not giving gifts to the public: he is, rather, selling goods to the public, and he needs the money because, unlike his professorial counterparts, his salary is generally not underwritten by a large, benevolent institution.

Libraries and universities pose an apparent threat to copyright guardians because they ostensibly represent a gift economy; they exist to disseminate information for the betterment of the commonweal, rather than to control the flow of information for the benefit of creators. The concept of intellectual property is a cornerstone of the market economy. Its adherents view with suspicion anyone who fails to share their free-market framework for cultural creation, and thus easily classify dissenters as pirates or thieves.

The goal of all copyright laws is to restrict cultural gift-giving: to ease the transformation of facts into artifacts, to give creators a vast supply of velvet rope with which to regulate access to their creations. This impulse is not wholly negative; for one thing, artifacts can popularize facts. But the ultimate text of the law that became the

Copyright Act of 1976 removed any residual doubt about exactly whose interests copyright was designed to protect. "In virtually all the contested points in the bill it seems that in the final revision all were reconciled in behalf of copyright owners. Is that not essentially true?" Representative Robert W. Kastenmeier observed when the bill was first presented to Congress in 1965. Eleven years later, his observation was still valid: the final bill was tailored to give creators and their corporate allies every possible opportunity to control and profit from their creations. The law expanded the standard copyright term from twenty-eight years, renewable once, to fifty years after the death of the author. It explicitly granted copyright protection to a range of new works. In fact, it guaranteed that any form of creative expression that originated in the United States would enjoy copyright protection from the moment of its creation, whether or not it had officially been registered.

The concurrent development of the Copyright Act of 1976 and the automated library illustrates two opposing views on new technologies, intellectual property, and the implications of the former for the latter. The same technologies that inspire some people to dream of a better world can inspire apocalyptic visions in others. Meeting the challenge to greatness depends entirely on how "greatness" is defined—and on who is doing the defining.

In 1965, with the Intrex- and UNIVAC-powered futures apparently imminent, an information scientist named Watson Davis spoke to the National Microfilm Association about something he called "the universal brain": a mechanized library that promised to store and organize the sum of human knowledge. The world already had the technology to build the universal brain, Davis enthused; social friction seemed the primary impediment to its construction. "Is cooperation between librarians and organizations so difficult that 'one big library' can not be accomplished?" Davis asked his audience. "A little organization, cheerful argument, and gentle pressure from users and financial supporters such as government and foundations would make it possible. The know-how exists and we only need the let's-do."[62]

The let's-do has always been in short supply. If computing technology is indeed the centerpiece of seductive dreams—a license for optimists to imagine idealized futures—then one of the most persistent fantasies is the cheerful abdication of analog attitudes in favor of some impending digital utopia. In fact, technologies that promise to solve the world's problems often end up entrenching them further. In the 1960s and 1970s, the world stood on the verge of a new informational era. But its advent portended the demise of the old one—and an incumbent rarely welcomes his challenger.

As for Project Intrex, it faced a dearth of let's-do and, ultimately, never came together. "The project seemed to be a bottomless financial pit," wrote the historian Colin Burke, perhaps the world's foremost Intrex authority.[63] Burke also noted that the "technology was not ready to provide the high-powered information engine their goals demanded. Those that put too much faith in rapid technological advances, like Intrex, had to spend too much time waiting for the technology to appear." The project collapsed by 1972, with little to show for itself except "two special 'combined' terminals that had already become somewhat outdated and, perhaps, some students and staff who were motivated to continue the search for the automated library."[64] As Intrex disintegrated and the Copyright Act of 1976 materialized, the organizers of both were only dimly aware that they would soon have to deal with the biggest library of them all: the Internet.

4

THE INFINITE LIBRARIAN

I n the waning months of 1970, an obstreperous drifter named Michael Hart returned to his hometown for good.[1] Like many of his generation, the twenty-three-year-old Hart wanted to do something meaningful with his life; like many of his generation, he wasn't exactly sure what that something would be. "People are shortsighted + shallow + world is suffering as a result," he scribbled on a sheet of paper. "A. Don't know if anything can be done about it. B. Find out."[2]

Since leaving the Army in 1968 he had chased those answers, with little to show for his efforts.[3] He spent months wandering America's interstate highways as a cross-country hitchhiker, but the road brought him no great epiphanies.[4] For a while he had attempted to effect social change as a singer of didactic folk songs. But "the World at large wasn't listening," Hart later wrote, "and I finally decided that if Dylan, Christ, and Simon & Garfunkel couldn't have the effect I wanted then neither would I be able to do it in that manner."[5]

You can say this for Michael Hart, at least: he didn't expect to succeed where Jesus Christ had failed. But that was virtually the only concession he was willing to make to his own limitations. Since childhood, Hart had been convinced of his unique greatness. "I wonder at those times whether he thinks of himself as some kind of superman who is above and beyond the level of the rest of us poor mortals," his mother wrote to family friends when Michael was nineteen.[6] As an Eagle Scout candidate in 1964, the young Hart had surprised the members of his review board by warning them not to mistake him for an ordinary civic-minded teenager. "I told them," he recalled, that "I was revolutionary."[7]

And Hart was indeed revolutionary, if you loosely define *revolutionary* as "disruptive." Though these sorts of things aren't really tracked by any official body, Hart was likely among the least compliant teenagers in America during the 1960s. As a boy, on his first day at a new school in central Illinois, Hart announced his arrival by threatening to sue the administration over its dress code.[8] Twice he enrolled at the University of Illinois at Urbana-Champaign; twice he withdrew in pique, enraged at perceived slights from instructors who were unable or unwilling to deal with his temperamental genius.[9] ("My Integrity was being lost to the System," he wrote later, by way of explanation.)[10] Drafted into the military in 1966, he immediately antagonized his commanders with a refusal to swear allegiance and faith to the Constitution.[11] "I can beat the Army," he wrote to his family.[12] He couldn't—and he didn't—but he kept on trying, to the extent that his eventual discharge from active service in 1968 surely prompted several champagne toasts somewhere in the bowels of the Pentagon.

It wasn't that Hart was a peacenik, not particularly. (Though he sought conscientious objector status, his mother suspected that this tactic may have been a ploy to get out of the Army.)[13] It was more that Hart couldn't stand being made to march. "I just wasn't brought up to blindly obey anyone, particularly a stranger, no matter how authoritarian," Hart observed.[14] Professors, administrators, superior officers—Hart bristled at anyone who demanded intellectual deference, anyone who used "because I said so" as a rhetorical trump. "I did not understand that a person could become attached to a method of doing things, would hold on to that method, fight for it, try to destroy any other in an effort to avoid having to learn something new," Hart later wrote.[15] He defined himself by his willingness to reach his own conclusions, and to defend them in the face of universal disagreement. His personal papers include the transcribed lyrics to two thematically relevant contemporary pop songs: "Driftin'" and "Different Drum."[16]

Hart traveled to his own beat as a matter of principle. A basic problem with society, in Hart's estimation, was that too many of its citizens were stuck in their own silos, accepting received wisdom as dogma, unwilling or unable to think for themselves. Hart hadn't had much success convincing others to open their minds, however, and he wondered if his window of opportunity was closing. "At the end of the 60's the Age of Apathy began and I started losing interest in my audiences because ther [sic] were losing interest in everything," he wrote, explaining why he had abandoned his musical ambitions.[17] So he returned to Urbana-Champaign in the winter of 1970, in search of a third shot at a college degree. There he learned that the instrument with which he would effect change wasn't the guitar after all. It was the keyboard.

HART matriculated (again) at the University of Illinois at Urbana-Champaign in 1971, determined to get good grades "in order to shove it up the ass of society," as if the most subversive thing he could do, as a man from whom nothing was expected, was to show that he was the most capable person around.[18] The plan made sense to Hart, at least, and for perhaps the first and last time in his life he decided to succeed on someone else's terms. He enrolled in a program called Individual Plans of Study, which allowed motivated students to design their own majors and chase their own interests, wherever those interests might lead them.[19] He ended up at the U of I's Materials Research Lab.

At the time, the lab housed a large Xerox Sigma V mainframe

computer, a $300,000 leviathan that barely resembled modern-day machines.[20] The room-size behemoth was covered in switches and buttons and lights; it screamed *Science!* and, indeed, like other computers of the era, the Xerox Sigma V was intended for use by science and engineering departments. The very word *computer*, first used in 1613, originally referred to specific individuals tasked with performing calculations.[21] A similarly narrow arithmetical lens informed the view of contemporary computers in the mid–Space Age. "We are still not thinking of the computer as anything but a myriad of clerks or assistants in one convenient console," wrote Columbia University professor Louis T. Milic in the inaugural issue of *Computers and the Humanities* in 1966. "We do not yet understand the true nature of the computer. And we have not yet begun to think in ways appropriate to the nature of this machine."[22]

Five years later, Milic's observation remained valid. To the materials scientists of the University of Illinois in 1971, it perhaps wasn't obvious that the monolithic Sigma V *had* a nature, much less one that was worth learning. They were content to entrust its daily operations to several young system programmers, who acted as data chauffeurs, guiding the computer to the researchers' desired destinations while ensuring it didn't break down or catch fire. When not running programs for lab personnel, the system operators enjoyed free rein on the machine and were encouraged to spend that unsupervised time exploring its capabilities.

Hart befriended two of the system programmers and started hanging around the lab, which was cool and quiet and conducive to study. On July 4, 1971, rather than brave the summer heat for a long walk home, Hart decided to spend the night in the mainframe room, so he ran out for groceries and settled in for history's nerdiest slumber party.[23] The system programmer on duty was a friend of Hart's brother, and, for some reason, he decided that night to give Hart his very own account on the Xerox Sigma V—which, Hart later claimed, may have made him the first private citizen on the Internet.[24]

Hart realized that he had been given great power, and this power made him hungry. Reaching into his grocery bag for some food, he noticed that a checkout clerk had included, free of charge, a pamphlet containing reproductions of various patriotic documents.[25] The insert began with the Declaration of Independence—America's foundational text, the document that announced a revolution so significant that the entire world took notice. For a young man keen on kindling an epochal revolution of his own, the symbolism was surely too obvious to miss.

In that moment, Michael Hart saw the future. Not two hours after obtaining an account on the Xerox Sigma V, Hart later bragged, he had divined the machine's core function. He announced that "the greatest value created by computers would not be computing, but would be the storage, retrieval, and searching of what was stored in our libraries."[26] And he quickly realized that *someone* would have to stock that library's digital shelves.

So, on that Independence Day, Hart resolved to transcribe the Declaration of Independence and put it onto the Xerox Sigma V for any and all to read. Using a Teletype terminal,[27] Hart typed it up in capital letters—computers did not yet support lowercase text[28]—saved the document on a hard-drive pack, and informed the other network users that the Declaration of Independence was now available in computerized format.[29] It was the first e-book.

Years later, Hart would burnish this act into legend—the genesis of a movement that would eventually spread across the world, one that would "undoubtedly become the greatest advance to human civilization and society since the invention of writing itself."[30] At the time, though, it seemed less like an opening salvo than a misfire; just another unnoticed folk song. (According to Hart, the Declaration was accessed only six times.)[31] The upload had its most profound effect on the uploader himself, Michael Hart, who was convinced that he had hit on something big, even if, or perhaps *because*, no one else shared his optimism.

The story of the modern free culture movement essentially begins here, in the early days of digital computing, on the margins of mainstream consciousness; its first protagonists unsupervised misfits such as Michael Hart who accomplished all they did simply because there was nobody around to stop them. Like many of his type, Hart believed that open information was intrinsically good, while rarely bothering to specify just how and why open information would change the world. To him, the progression seemed obvious. You couldn't change the world without changing minds, and you couldn't transform people's thought processes without giving them something new to think *about*. The powerful wanted to keep the masses ignorant and pliable by making information expensive and scarce. Hart concluded that digital networks could be used to set that data free.

Over the next forty years—intermittently at first, then consistently after the introduction of the World Wide Web in 1991—in the face of doubt, poverty, and general public indifference, Hart devoted himself to typing public-domain texts into computers for the benefit of the wider world. At first, he labored alone; later he managed a corps of volunteers and served readers from all inhabited continents. By the time of his death in 2011, Hart had helped digitize more than thirty-seven thousand books and historical documents—the Bible, the complete works of Shakespeare, *Areopagitica*, *Paradise Lost*, *The Federalist Papers*, *The Book of Mormon*, *The Pickwick Papers*, *Moby-Dick*, *O Pioneers!*, and scores of lesser-known works—as part of this earliest and perhaps most pure attempt to create and populate a digital library of the future.[32]

Michael Hart may not be the most important member of the early open-information movement. He is certainly not the only person who worked to digitize and disseminate computerized texts. But his long persistence, with near-monomaniacal focus despite little positive reinforcement, makes him an emblematic figure, and perhaps an instructive one, too. Just as the invention of the Gutenberg press had helped catalyze the Protestant Reformation, Hart thought that computerized texts would have a similarly metamorphic effect on modern society, that they would "break down the bars of ignorance and illiteracy."[33] E-books, he believed, would change the world—as soon as the world learned that they existed.

"You have to realize that for the first 17 years absolutely NO ONE paid any attention," Hart wrote to an online mailing list in 2006. "Even the friends that helped me thot [sic] I was nuts, and that it would never work."[34] But even Hart's enemies acknowledged his sincerity and dedication. Hart spent much of his adult life surviving on income he received from renting out rooms in his cluttered house to college students and forgoing traditional employment to spend his days and nights apostolizing e-books from his basement office. His walls were adorned with index cards onto which he had inked various aphorisms in block lettering with a thick black marker. Marcus Aurelius: A PERSON CAN NOT LEARN WHAT S/HE ALREADY KNOWS OR THINKS S/HE KNOWS. H. G. Wells: ULTIMATE COURAGE IS BRAVERY OF THE MIND. William Blake: I MUST INVENT MY OWN SYSTEM OR BE ENSLAVED BY OTHER MEN'S.[35]

For Michael Hart, the digitization of public-domain literature was a vocation in its original sense: a spiritual calling to a movement. But movements need names, and Hart delayed for a long while—for seventeen years following his transcription of the Declaration of Independence, to be exact—before choosing one. Lying on a mattress on the floor of his house in Urbana, Illinois, in 1988,[36] Hart pondered various names for his endeavor—The Electronic Book Factory, The Never Ending Library, Library Galactica[37]—before finally adopting one that elegantly captured both his own revolutionary ambition and his endeavor's transformative potential: Project Gutenberg.

MEN like Michael Hart are naturally drawn to libraries. For those who are intrinsically suspicious of authority and received wisdom;

for those whose curiosity transcends the boundaries of grade levels and lesson plans; for those who are fascinated more by process than by outcome, libraries can be a godsend. They are repositories of unfiltered information, and guarantors of America's autodidactic ideal. As Herbert Putnam wrote in 1895, a clever person who cannot or will not attend college "is still by familiar experience equipped to pursue a higher education in the public library: all the best that mankind has had to say in history, science, or art, being placed there freely at his service."[38] When Michael Hart began his life's work, that intellectual bounty was more plentiful and accessible than ever before.

If libraries nurtured dreamers like Michael Hart, then digital computers and networks activated them. "When I first learned I had access to the Internet back in 1971 it was as if one of those lights you see in the comics went off with a flash right over my head," Hart recalled.[39] Generations of disaffected nerds have felt the same way. The real world favors the handsome and coordinated, rewards conformism and compliance, and reinforces systemic inequities by design. The Internet offers an alternative. Thanks to its decentralized architecture, the network inverts the way that power and expertise tend to consolidate in the hands of big organizations that can use that power to perpetuate the divide between rich and poor. "With the advent of personal computers and with information becoming more and more the direct cause of production, we have a great chance at a new renaissance," Hart observed in the 1980s.[40] For many people, the Internet and computers appeared to be tools that would rebalance the world.

The American government didn't realize it was underwriting something so radical. One of the results of the post-*Sputnik* surge in science spending was the creation of a Department of Defense office called the Advanced Research Projects Agency, or ARPA. Initially responsible for working on spaceflight, ARPA soon shifted priorities and became the government's long-term research-and-development laboratory, specializing in projects with minimal short-term applications that might nevertheless prove important down the line. One of these projects was the distributed computer network that would eventually become known as the ARPANET.

The ARPANET was designed to let ARPA-affiliated computers talk to one another and transmit data remotely. But J. C. R. "Lick" Licklider, an early executive at ARPA, also harbored bigger plans for the network. Whereas the Department of Defense hoped the ARPANET would make America stronger and safer, Licklider hoped it would make humans smarter and happier. "The idea on which Lick's worldview pivoted," according to Katie Hafner and Matthew Lyon in *Where Wizards Stay Up Late*, "was that technological progress would save humanity."[41] Machines' storage, retrieval, and processing capabilities would help their human operators to be more productive by minimizing error and mundane repetition. Humans would, in turn, improve and refine the machines. This cycle of mutual benefit and development would continue until humanity arrived at what Licklider described as a "man-computer symbiosis."

Licklider had attended the Project Intrex conference and was enthusiastic about the prospect of digital libraries. "We need to substitute for the book a device that will make it easy to transmit information without transporting material," he wrote in a 1965 book titled *Libraries of the Future*.[42] The ARPANET took a step in that direction. The network would support a new, flat system of information sharing, an information commons with multiple points of entry. Licklider saw the "intergalactic network," as he called it, as his gift to the world, one that would keep on giving. Though he left ARPA long before the ARPANET was completed, his spirit and ambition guided his successors' work.

The early ARPANET consisted almost entirely of academic affiliates using institutional computers. The US government provided the network infrastructure and initially decided to restrict commercial use of the service. (The last of these restrictions wasn't lifted until 1995.) Thus these early users, prohibited from exploiting the network for profit, used it instead to foster the free exchange of information.

This munificent ideology was encoded into what the author Steven Levy described in his insightful book *Hackers* as the "hacker ethic." Hackers—a term for early computer programmers—wrote computer code and believed that other hackers should share their code and computing resources with their peers. This policy was, in part, a pragmatic one: at the time, computing resources were scarce, and possessiveness impeded productivity. But the attitude was also a conscious philosophical choice, a statement that the world ought to be open, efficient, and collaborative. Nowhere was this hacker ethic taken more seriously than at the Artificial Intelligence (AI) Lab at the Massachusetts Institute of Technology.

As Levy tells it in *Hackers*, the AI Lab was something of a socialist utopia for computer programmers. The lab was charged with developing thinking machines, and it deployed a lot of computers toward that end. These computers were programmed and otherwise tended to by a crew of young hackers, some of whom were paid employees, others just enthusiasts who congregated at the lab to be close to the objects of their desire. The hackers wrote code and maintained the lab's computers with radical transparency. Any hacker could access another's files to study or improve them. The computer terminals had no passwords; few locks were on the doors. For many of the hackers, the lab functioned as a surrogate dormitory. In 1986, a former AI Lab hacker named Richard Stallman recalled how he and other programmers would "stay up as long as you can hacking, because you just don't want to stop. And then when you're completely exhausted, you climb over to the nearest soft horizontal surface."[43] The next day, they would do it all over again.[44]

An intense, mischievous man who sported a dark beard, studied folk dancing, and sometimes wore a button reading IMPEACH

GOD,[45] Stallman was dubbed "The Last of the True Hackers" by Levy. He began his MIT career in 1974 as a graduate physics student, but spent his evenings working on computers at the AI Lab. After a while, Stallman examined his priorities and realized that they no longer included the study of physics; he dropped out of school in 1975 and embraced his digital destiny. Stallman became a full-time AI Lab programmer. Eventually, he emerged as the conscience of the hacking community, an unreconstructed champion of open information.

The same independent, inclusive ethos that empowered Stallman and the AI Lab hackers also created opportunities for people like Michael Hart—difficult, ambitious individuals who fared poorly in hierarchical systems but thrived in nonconformist environments. Like Stallman, Hart used computers to build the sort of world that *he* wanted to see, one in which knowledge was freely given rather than sold, one predicated on altruism rather than profit. "Idealism—the way you think + feel things ought to be—is soooooo wonderful!!!!!!" Hart once scribbled in his diary.[46] The early digital utopians saw computers as instruments of social realignment, the vanguard of a revolution that would remake the world along more egalitarian lines.

It didn't last. In the early 1980s, at the dawn of the personal-computing era, some of the AI Lab's hackers defected to a company called Symbolics, where they constructed commercial versions of the machines they once built at MIT. Others went to a competing firm. The AI Lab hacking corps dwindled, and its motivating principles came to seem increasingly obsolete. For Stallman, these defections only reinforced his commitment to free software.

Stallman believed that proprietary software was inimical to the hacker ethic, and that it impeded the free flow of knowledge. Treating software users strictly as customers rather than potential collaborators implied that the public could contribute nothing of value but money. "The rule made by the owners of proprietary software was, 'If you share with your neighbor, you are a pirate. If you want any changes, beg us to make them,' " Stallman wrote.[47] He couldn't abide this philosophy, so he quit MIT to develop a free computer-operating system. "With a free operating system, we could again have a community of cooperating hackers—and invite anyone to join," Stallman wrote. "And anyone would be able to use a computer without starting out by conspiring to deprive his or her friends."[48]

He called the program GNU, a "recursive acronym" that stands for "GNU's Not Unix." (Unix is a popular computer-operating system. A gnu is also a large, hairy wildebeest, an animal to which Stallman bears a faint resemblance, if you squint and use your imagination.) After working on the GNU Project for approximately two years, Stallman doubled down on his principles and founded a complementary organization called the Free Software Foundation.

The Free Software Foundation ("Don't think of free as in free beer; think free as in free speech," Stallman explained) was and is dedicated to the idea that unrestricted access to information and computer code is a simple matter of justice. The foundation argues that software ought to be liberated from licenses, passwords, and other restrictions that foster adversarial relationships among computer programmers and users. "My work on free software is motivated by an idealistic goal: spreading freedom and cooperation," Stallman wrote.[49] "I want to encourage free software to spread, replacing proprietary software that forbids cooperation, and thus make our society better."

Though Hart and Stallman worked in different fields, they shared similar tools, goals, and methods. They also shared similar blind spots. Free software and free literature by themselves will not necessarily improve society, and proprietary books and computer programs will not necessarily destroy it. The rise of a commercial trade around cultural artifacts can also end up expanding the audience for those artifacts, exposing them to people who might never have known they existed.

But commercial culture has never been particularly compatible with Hart and Stallman's brand of single-minded integrity. Free software, free e-books, free culture all promote user independence over reflexive compliance with authority and encourage users to become more than just passive recipients of other people's cultural products. "The perspective is that of changing the world from the bottom rather than changing things from the top down," Hart reflected. "If I had waited for those at the top to approve of what I was planning to do, even waited HALF as long as they wanted, then you would [have] never heard of me or of Project Gutenberg."[50]

THOUGH Hart liked to say that Project Gutenberg has existed continuously since 1971, in truth, it lived mostly in the traps of his mind for those first seventeen years. Hart's papers from that period make no mention of Project Gutenberg or any similar project—and the man seemingly recorded every stray thought he ever had. He continued to digitize famous texts—the Bill of Rights in 1972, the US Constitution in 1973, other patriotic documents in subsequent years.[51] He spent the 1980s digitizing the entire King James Bible. (It is a very long book.) But for most of this period, Project Gutenberg, to the extent it existed at all, was a hobby, intermittently pursued, fated for obscurity.

"My unhappiness is that of an unfulfilled potential," Hart wrote of himself in 1983.[53] He pursued employment in various technical fields: stereo sales, systems analysis, computer consulting. He became a prolific writer of complaint letters to businesses, institutions, and individuals who failed to meet his often impossibly high standards. ("I do not know what you want from Coleman," a representative of the Coleman outdoor products company wrote in plaintive response to one of Hart's letters.)[53] He briefly considered dropping out of society to found a self-sufficient farm, "complete with power generators, flour mill, + maybe a loom if anyone can run it."[54] He got married. The union didn't last. All the while, he was wondering how and when he would make his mark.

In 1984, Apple released the Macintosh computer, one of the first home computers for the nontechnical user, and a market for Hart's product slowly began to emerge. The gargantuan mainframes of the 1960s had given way to smaller computers suitable to the home office, produced and marketed by entrepreneurs who realized the potential for consumer revenue. Microsoft's Bill Gates—whose MS-DOS and Windows operating systems eventually dominated the PC market—and Apple's Steve Jobs, for instance, were unmoved by the hackers' utopian rhetoric. They wanted to make money, not save the world. But by bringing computers to the casual user, their products also amplified the message of those technologists who were operating from nonmaterial motives.

Hart joined several clubs and organizations for personal-computing enthusiasts and met other people who shared his interests, if not his grand ambitions. In 1988, he had a premonition "that something very powerful and fulfilling was just over the horizon."[55] Encouraged by the ascendance of the personal computer, and buoyed by the personal connections he had made in the various clubs to which he belonged, he took the first steps toward turning his e-book hobby into Project Gutenberg. Hart gradually started floating his idea to the wider world. "Dear Word Perfect User Group Members," one of his pitches began. "How would **YOU** like to be part of the biggest book project since Gutenberg invented the printing press?"[56] Project Gutenberg, at the time, featured approximately ten texts. Few others shared Hart's optimism that it would ever contain many more.

"I am not convinced that your idea has any future to it. I find it difficult to see why anyone would want to sit in front of a computer screen and read a book," an official at Illinois Benedictine College, where Hart served as an unpaid adjunct professor, informed him in 1988, suggesting that he instead focus on "some way of animating the texts and turn them into some sort of video."[57] That same year, a friend in California told Hart that a professor at San José State had scoffed at the idea of a digital library. "Seems that librarians as well as publishers fear for their jobs if electronic publishing becomes popular," the friend observed. "Boo Hoo!"[58]

Hart admired Ayn Rand's novel *Atlas Shrugged*, which was loosely devoted to the idea that geniuses tend to be subsumed by the mediocrities that surround them and are usually unappreciated in their own time. But unlike Rand's protagonist, Hart was unwilling to forsake those who had spurned his labors. He clung to his mission of bringing e-books to the world, whether or not its inhabitants either wanted or deserved them. "I know I will never be happy without trying to see if I can change the world for the better in a major manner," he wrote in 1990.[59] That January, Hart traveled to the American Library Association's midwinter meeting to proselytize for e-books and Project Gutenberg. There, he vowed, "There will be 10,000 Machine-Readable-Texts available by Dec. 31, 2000, even if I had to make them all myself."[60]

IN 1990, a British computer scientist named Tim Berners-Lee wrote an article for a house newsletter at CERN, a particle-physics laboratory in Switzerland. Berners-Lee programmed software at CERN, and, like many idealistic coders before him, he had become enamored of the Gospel of Richard Stallman. "A source of much debate over recent years has been whether to write software in-house or buy it from commercial suppliers. Now, a third alternative is becoming significant in what some see as a revolution in software supply," he wrote, referring to the Free Software Foundation and the GNU Project. Berners-Lee wondered whether Stallman's ideas might not be applied to the work he was doing for CERN. "Just as we publish physics for free, should we not in certain cases 'publish' our software?" he asked.[61] One year after publishing his Stallman-inspired note in the CERN house journal, Berners-Lee followed through on that note's central idea when he released to the world, for free, a project that he called the World Wide Web.

Though the World Wide Web is often used as a synonym for the Internet, they are not the same thing. The "information superhighway" metaphor may be hoary, but it still has its uses. A highway is just a road, designed to carry all sorts of vehicles without discrimination. Automobiles, trucks, motorcycles, trailers—they all use the highway, and the highway doesn't favor one over another. The road is simply infrastructure. You can think of the Internet as a highway that is used by different programs. Your e-mail service is one of them. Instant-messaging applications are another. The World Wide Web is just another of these programs: a piece of software that allows computers to talk to one another.

The Web became popular because of its linking capacity. In his proposal for Memex, Vannevar Bush advanced the idea that the associative trails between two disparate thoughts or facts could be captured and stored. The Web put a version of this idea into practice by allowing its users to link directly to other documents or websites: a feature called hypertext. The World Wide Web was modeled after an actual web, composed of threads—hypertext links—that spun out in all directions, connecting various far-flung nodes, or websites.

Like its arachnoid namesake, the Web was good at drawing others in. As more and more people dialed online—especially after the 1993 release of Mosaic, the first visual Web browser—Project Gutenberg grew into an actual project involving more than just Michael Hart and his increasingly weary fingers. As word of the initiative spread, volunteers materialized. "Your project sounds like a wonderful service to mankind and will no doubt increase in value," wrote a Houston, Texas, man in a 1992 letter to Hart. "How may a person support your efforts?"[62]

"Project Gutenberg, what can I say. I wish I could support it with money," wrote another supporter in 1993, enclosing a floppy disk containing a hand-digitized book. "Instead, I offer public domain etexts in support of this splendid idea."[63]

"Hello, Michael," wrote a Watertown, Massachusetts, man in 1995. "This diskette includes all the files for the HTML version of *A Heap o' Livin'* by Edgar A. Guest."[64]

"Project Gutenberg is currently posting four books per month in what is called Plain Vanilla ASCII, meaning the files can be read easily on virtually all computers and programs," Hart wrote in March 1993.[65] Two years later, that rate of production had tripled. As the online archive grew, Project Gutenberg began to attract media attention. Even though some journalists did not quite see the point of the digital library, they were nevertheless struck by its novelty. "Why exactly you'd want to download the entire text of Mark Twain's *A Connecticut Yankee in King Arthur's Court* is unclear," an Associated Press reporter wondered in February 1995, "but there's something wonderful about the idea that it's just sitting on its virtual shelf, waiting for you."[66]

Project Gutenberg maintained an online mailing list for volunteers and well-wishers, and at the end of every month, Hart issued a newsletter listing the titles that had recently been digitized. In the March 1995 Project Gutenberg newsletter, he proudly announced the library's 250th text: a rambling, occasionally bizarre, and incredibly sincere e-book called *A Brief History of the Internet*, written by none other than Michael Hart himself.[67] The book began with a reflection on the history of Project Gutenberg and how far it had come from its obscure origins: "Today there are about 500 volunteers at Project Gutenberg and they are spread all over the globe, from people doing their favorite book then never being heard from again, to PhD's, department heads, vice-presidents, and lawyers who do reams of copyright research, and some who have done in excess of 20 Etexts pretty much by themselves; appreciate is too small a word for how Michael feels about these, and tears would be the only appropriate gesture."

In the early 1980s, Hart had predicted that computing technology might stimulate a new Renaissance. At the time, it was just another of his windy and implausible pronouncements, more empty rhetoric from an embittered dilettante librarian. But the recent success of Project Gutenberg had given Hart new hope. A digital Renaissance still might not have been *imminent*, but the prospect of a new cultural flowering was no longer just a manic delusion. With two hundred fifty e-books behind him and many more ahead, Hart was more confident than ever that his long-delayed revolution was at hand. "Michael Hart is trying to change Human Nature," he wrote of himself in *A Brief History*. "He says Human Nature is all that is stopping the Internet from saving the world."[68]

IN 1995, Congress noticed that it had a copyright problem. During the nineteen years since the Copyright Act of 1976 had been signed into law, many European nations had amended their copyright statutes to give creators control over their works for up to seventy years after their deaths. American law, by contrast, still limited the posthumous copyright term to a mere five decades. To address this discrepancy, Senator Orrin Hatch introduced a bill called the Copyright Term Extension Act. Under the act, new works in America would remain under copyright until seventy years after the creator's death, thus bringing American copyright laws in line with international norms. But when copyright terms expand, the public domain contracts. In 1995, the Senate Judiciary Committee held hearings to determine whether the trade-off was worthwhile.

The most popular answer (among those prominent enough to be called to speak) was *yes*. In grand congressional tradition, a long line of celebrities submitted testimony arguing for a copyright term extension. Alan Menken, a composer who had scored animated films such as *Beauty and the Beast* and *The Little Mermaid*, announced that leaving the American copyright term in its current state would be "unjustifiable to all Americans, particularly at a time when we are positioning ourselves as a world leader on the global information superhighway."[69] Other musicians entered prepared statements into the record reiterating the injustice of the current situation.

"When I began my career as a songwriter, I believed that I was building a business that would not only bring enjoyment to people throughout the world, but would also give my children a secure base from which they could, in turn, build their own lives," wrote Carlos Santana.[70] "The impression given to me was that a composer's songs would remain in his or her family and that they would, one day, be the property of the children and their grandchildren after them," wrote Bob Dylan.[71] "On a daily basis, I wear many hats," wrote Don Henley, but instead of expanding on that intriguing image and describing the origins of his capelophilia, he simply explained how unjust it would be if, in 2050, his grandchildren were unable to collect the royalties from "Tequila Sunrise."

As the songwriters argued for a copyright term extension, others attempted to diminish the public domain. Jack Valenti, president of the Motion Picture Association of America (MPAA) and a longtime copyright lobbyist, portrayed the public domain as the equivalent of a weedy, untended vacant lot, telling Congress that it was "important to understand that public domain means nobody really cares because nobody owns it."[72] After Valenti concluded his testimony, the committee chairman, Senator Hatch, noted exactly how much he enjoyed being able to view first-run movies at the offices of the MPAA, and said that "the big reason I like to do it is, frankly, just to be able to say hello to you on a regular basis."[73]

A few law professors defended the value of the public domain. The American University professor Peter Jaszi noted, "In my academic wanderings through the legislative history of American copyright, I have been struck by how seldom and how little the Congress has heard from the users of public domain material, a loose community that is both more numerous and more diverse than one might expect."[74] The sentiment of the room, however, was not in Jaszi's favor.

What was the purpose of the public domain? Congress hadn't cared to seriously consider the question in 1976, and the body remained indifferent in 1995. "It should be obvious that if laws such as [the Copyright Term Extension Act] continued to be passed every 20 years or so, that nothing will ever enter into Public Domain status again and the work of people such as the Internet Wiretap, the Online Book Initiative, and Project Gutenberg will soon be over," Hart wrote to supporters at the time.[75] Just as the AI Lab exodus in the early 1980s had threatened the survival of the hacker ethic, the popularization of the Web in the mid-1990s—and the concurrent emergence of its commercial potential—threatened to marginalize the digital utopians who had been its first colonists.

"While I certainly lay no claims to inventing the Internet, I was the first I have ever heard of to understand what it was to become over the first few decades of its existence," Hart wrote on his blog. "This pioneering spirit is usually one of the first things to go—once the 'dude' and 'suits' have their way and starting with the politicking that places people in power who have no idea of the who, what, where, how, why and when power originated, the power they have usurped from those who created it."[76] For Project Gutenberg, this point was reinforced early in 1996.

The University of Illinois at Urbana-Champaign had indirectly powered Project Gutenberg since its rudimentary mainframe days back in the 1970s; after he graduated, Hart had been allowed continued access to the school's computing resources. Not only was Hart an alumnus, he also worked for the university as a consultant, was a familiar figure on campus, and had come to count on the school's goodwill and support, both of which appeared to be cheerfully provided. In 1989, an official in the university's computing services office wrote a letter of endorsement for Hart and his nascent digital library.

"Were our resources greater, we would take on Project Gutenburg [*sic*] ourselves," the official wrote. "As it is, we can offer limited assistance and unlimited encouragement."[77]

Just over six years later, the school's assistance and advocacy for Project Gutenberg had declined. The needs of the institution were changing. Subscription database companies that charged access fees and reinvested that revenue in improved services were eclipsing free services such as Project Gutenberg, with their unadorned text files and their volunteer laborers and their intensely antistatist founders. The first online library was no longer necessarily the best online library.

When Hart first encountered the Xerox Sigma V in 1971, the Materials Research Lab had encouraged its system programmers to experiment, in hopes that they might find new uses for the machines. Now the university knew what computers were for—and what they weren't for. Project Gutenberg provided insufficient value to the university to merit continued institutional support. In January 1996, citing increased demand for its computing resources, the University of Illinois's Computer and Communications Services Office informed Hart that his access to the university's computers would be revoked that coming July, almost twenty-five years to the day that Project Gutenberg began.[78]

Hart was stunned. Project Gutenberg had been growing steadily, and Hart's goal of putting ten thousand e-books online by the millennium seemed difficult but attainable. Eviction from the University of Illinois's system would devastate that ambition. He realized that the school had to be convinced to reverse its decision. He had to show it just how much Project Gutenberg was really worth.

So Hart asked fans and friends of Project Gutenberg to e-mail the university and urge a reconsideration. "Please keep [your e-mails] short, just say that you want to support this project and you think this is a worthwhile use of the computer systems at the University of Illinois," Hart wrote. "It doesn't really cost much of anything to keep us on the system, it is that the political climate here is such that we are in danger of having our accounts thrown out with the general cleanup of all people without the proper official connections; something we have had very little of during our 25 years of service, and which would not be easy to get."[79]

Hundreds and hundreds of e-mails arrived from around the world, testifying to just how much Hart's homemade library meant to its many users. "I feel that Project Gutenberg, the GNU project and similar efforts are really the embodiment of the spirit of the Internet. These groups do their work as a great labor of love, as a gift to all of humanity," wrote one supporter.[80] "I think that it would be a terrible shame if during this time of enormous growth of the Internet, Project Gutenberg turned into one of those things that *used to be around*," wrote another.[81]

"When people have fast, convenient access to works of literature large and small, important and insignificant, they have access to a broader range of thought than they might otherwise have. And they have the opportunity to compare, contrast and synthesize the works into a body of knowledge that they could not reasonably have been able to amass without such access," wrote yet another Gutenberg fan. "Please continue Project Gutenberg. It's [sic] cost is so much less than its value."[82]

The inflood of support set the stage for a classic narrative denouement in which the underdog triumphs, the villains repent, and the heroes all live on the network happily ever after. But real life rarely resolves as neatly as fiction. The e-mail campaign failed to persuade the university to keep Project Gutenberg alive.

"We are not sure how we will be reaching you after midnight, but we will be working on as many ways as possible," Hart wrote to Project Gutenberg supporters that June, one day before his system access was scheduled to expire[83]:

> For all history there has never been enough of anything for us except a supply of air for us to breathe, and now, for the first time,

there has been enough copies of some books, that everyone for the rest of history can have a copy.

Please don't let this first time for something like this be the last.

While the eviction was a setback for Project Gutenberg, it didn't destroy it forever. In 1997, the project found a new home at Carnegie Mellon University. Eventually, Project Gutenberg incorporated as a nonprofit foundation. All the while, Hart and his volunteers kept at their lonely digital labors.

On September 4, 1997, Hart triumphantly announced the release of the thousandth e-book in the Gutenberg archives, Dante's *Divine Comedy*. Though he took a moment to celebrate what he deemed "the pinnacle of my career," he couldn't help but predict that future milestones would likely be few and far between, unless the American reading public took control of its nation's copyright laws. Project Gutenberg had become an eloquent counterargument to copyright advocates' dismissive claims about the public domain. It demonstrated just how easily a network could be used to breathe new life into classics that might otherwise go unseen.

Despite the existence of initiatives such as Project Gutenberg, despite the emergence of the Internet as a new medium for information retrieval and distribution, the same official attitudes about intellectual property prevailed. The public domain was regarded as a penalty rather than as an opportunity. Parochial concerns were conflated with the public interest. The rise of the Internet might portend an informational revolution, but from the standpoint of the people in power, Hart warned, revolution was a *bad* thing.

"Every single time a new publishing technique has promised to get the common people a home library, laws have been passed to stop, dead in its tracks, this kind of 'Information Age,'" Hart wrote. "Information Age??? For Whom??? We hope it will be for you."[84]

5

THE CASE FOR THE PUBLIC DOMAIN

In 1995, a New Hampshire man named Eric Eldred sat down with his three young daughters to help them with some schoolwork.[1] The Eldred triplets had been assigned to read Nathaniel Hawthorne's *Scarlet Letter*, and they found it slow going. Eldred wondered if perhaps Hawthorne's prose wasn't the problem so much as the outdated bound format in which that prose was presented. So he created a website and put *The Scarlet Letter* online, studded with explanatory hyperlinks: a bedazzled, annotated version of the plain-text e-books found at Project Gutenberg.

While the project didn't turn his daughters into Hawthorne fans, Eldred enjoyed the work all the same, and he started to digitize Hawthorne's other novels. Soon, he had built a comprehensive online archive of the author's entire bibliography. Such a project might not seem particularly novel from today's vantage, but in the mid-1990s, Eldred's efforts were absolutely ahead of their time. In 1997, a National Endowment for the Humanities initiative called edSITEment deemed Eldred's Hawthorne site one of the twenty best humanities resources online.[2]

Eldred uploaded more works to his "Eldritch Press" website: Henry James's *Daisy Miller*, Louisa May Alcott's story "An

Old-Fashioned Thanksgiving," several books by William Dean Howells. ("Copyright should be perpetual," Howells had argued in 1879.)[3] Eldred's project began to assume a broader significance. "I imagined that what I was doing was promoting democracy, respect for other people, mutual understanding, literacy, appreciation for literature . . . and a lot of nice warm and fuzzy things," Eldred would later write. "I wanted to show everyone how great all this was."[4]

Like Project Gutenberg in its early years, Eldritch Press was a one-man operation; the texts were created on Eldred's home computer and uploaded over a cable modem connection. By 1998, despite these limitations, Eldritch Press featured over a hundred megabytes' worth of classic titles formatted for easy online reading.[5] The bulk of these titles had first been published before 1923, and thus, in 1998, stood unquestionably within the public domain.[6]

Eldred lived in Derry, New Hampshire, a large farm town near Manchester that was, for a time, the home of poet Robert Frost. It was Eldred's eventual hope to honor the poet by creating an Eldritch Press edition of *New Hampshire*, Frost's Pulitzer Prize–winning 1923 poetry collection, which was scheduled to enter the public domain on January 1, 1999.[7] The book contains many of the poet's best-known compositions: "Stopping by Woods on a Snowy Evening," "Nothing Gold Can Stay," along with the title poem, in which the narrator begins by denouncing commerce as disgraceful.

Poets are expected to proclaim unprofitable sentiments from on high; corporations, however, do not generally share the artist's disdain for business dealings. In October 1998, partially at the behest of Disney and other companies terrified of losing the copyrights on their lucrative corporate mascots, Congress finally passed a version of the copyright term extension bill that Senator Orrin Hatch had first introduced in 1995.[8] The law was called the Sonny Bono Copyright Term Extension Act (CTEA), in honor of the recently deceased congressman and songwriter. Bono's widow, Representative Mary Bono, claimed that her husband had "wanted the term of copyright protection to last forever," and his former colleagues in Congress basically decided to grant his wish.[9] The CTEA extended the copyright term on any given work created before 1978 to ninety-five years after its first publication. Works published after 1977 would remain under copyright until seventy years after the author's death.

The bill was officially celebrated as a victory for artists and writers. "Extending the term of copyright protection by twenty years will ensure that the American public continues to enjoy the contributions made by our creative community," Congressman Bill McCollum, who represented Orlando, Florida, home of Walt Disney World, said upon the bill's passage in the House.[10] But it also put *New Hampshire* out of Eric Eldred's reach.

The CTEA was one measure in a series of congressional bids to modernize federal intellectual-property statutes. The day after he signed the CTEA into law, President Bill Clinton did the same for a bill called the Digital Millennium Copyright Act (DMCA), which fortified federal efforts to prevent unauthorized access to copyrighted material. Not only did the DMCA prohibit the circumvention of copy-protection technology—for example, the mechanisms that require you to activate an expensive computer program before using it—the law also forbade people from using the Internet to explain how to evade copy-protection, or even linking to those explanations.[11]

The year before, in December 1997, President Clinton had signed a kindred bill, the No Electronic Theft Act (NET Act), which stipulated felony criminal penalties for people charged with the unauthorized online distribution of copyrighted material, even in cases where the distributor had no profit motive. The NET Act was intended to close the so-called LaMacchia loophole.[12] In 1994, federal prosecutors indicted an MIT undergraduate named David LaMacchia for allegedly operating an online bulletin board that people used to freely trade copyrighted computer software. While it appeared to prosecutors that LaMacchia had helped facilitate software piracy, his actions were not considered criminal under the terms of the Copyright Act, since he had not acted for any clear commercial purpose. Though prosecutors instead charged LaMacchia with wire fraud, the case was eventually dismissed on the grounds that the wire-fraud charge was misapplied. Even so, Judge Richard Stearns sharply criticized LaMacchia in the dismissal, characterizing his alleged actions as at best "heedlessly irresponsible, and at worst as nihilistic, self-indulgent, and lacking in any fundamental sense of values."[13]

In a 1997 statement to the House Subcommittee on Courts and Intellectual Property, Register of Copyrights Marybeth Peters noted that, absent strong laws to deter cultural-chaos agents like LaMacchia, heedless hackers would continue to use the Internet to destroy legitimate markets for creative works. "While the existing 'commercial purpose' requirement, in the world of physical copies, has served to limit criminal liability to piracy on a commercial scale, a new standard is needed in the digital environment, where significant economic damage can be caused without a commercial purpose," Peters testified.[14] Information, after all, wants to be expensive.

These three bills gave copyright holders effective recourse against self-indulgent nihilists, amoral fraudsters, and other blithely piratical souls. But they also hampered the work of volunteer librarians such as Michael Hart and Eric Eldred. "The new laws make no sense when the technology of electronic publishing has made it so cheap and easy for a person like me to publish much better stuff than the print publishers can do," wrote Eldred upon passage of the CTEA. "They make sense only if they are means to retard technological change and serve as protectionist measures to allow the old print publishers to continue in their ways."[15]

"When I first started Project Gutenberg nearly 30 years ago, I predicted that by the time that we could put a million books in a single box we could hold in one hand, that they would make it illegal to do so," Hart wrote to Book People, an online mailing list for e-text devotees.[16] The list's name referred to the novel *Fahrenheit 451*, Ray

Bradbury's tale of a world in which books had been banned and were regularly burned by the state. Hart interpreted the new law as a similarly dystopian measure, an active attempt to serve corporate interests and check the spread of knowledge. On the Book People mailing list, Hart wrote, "It will take some SERIOUS revolutionary times and persons . . . if we are ever to get back the right of public domain as it is and was supposed to be."[17]

Eric Eldred agreed. The new law appeared to have ruinous consequences for his work with Eldritch Press. If he put Frost's book online without authorization, he risked criminal charges. Eldred had been living on workers' compensation payments ever since repetitive-stress injuries had forced him into early retirement, didn't want to go to jail, and couldn't afford to pay a fine. "After many years, and after I am dead, society will awake to the insanity of all this," he wrote in frustration to the Book People list.[18] In the meantime, he shuttered Eldritch Press for good. Then he got a phone call from a Harvard professor who was seeking a plaintiff for a test case to challenge the laws governing the public domain.[19]

Lawrence "Larry" Lessig was a law professor at Harvard University, where he was affiliated with the Berkman Center for Internet & Society. Though he had come to lead a movement looking to liberalize America's copyright laws, he didn't immediately seem like a particularly radical fellow. Lessig "grew up a right-wing lunatic Republican" and had clerked for the conservative jurists Richard Posner and Antonin Scalia.[20] He had a receding hairline and a memorably large forehead offset only by a small pair of eyeglasses. The *Los Angeles Times* once described him as "very pale and very quiet, as if he doesn't want to bother the fellow in the next cubicle."[21]

As Scalia's clerk, he helped to convince the Supreme Court that its archaic internal computer system was in desperate need of an update by demonstrating to Scalia the thesaurus function on a more modern machine.[22] Following that victory, he continued in his efforts to persuade the government that technological change was nothing to fear.

Lessig, like Eldred, believed that copyright in America had strayed from its original purpose, and that lawmakers failed to recognize the realities of digital culture. "To digitize a book is to copy it," he would later write. "To do that requires permission of the copyright owner. The same holds for music, film, and every other artifact of our culture protected by copyright. The effort to make these things available to history, or to researchers, or to those who just want to explore is now inhibited by a set of rules that were written for a radically different context."[23] Congress clearly had no intention of rebalancing those rules; therefore, Lessig concluded, he would have to convince the courts to strike down the recent copyright extension and revitalize the public domain.

Lessig initially identified Michael Hart as a good plaintiff for his CTEA test case and even flew to Urbana to discuss a potential alliance. But Hart insisted on taking an active role in the case and demanded that he be allowed to append a polemical personal statement to Lessig's legal briefs. Hart's manifesto "was great rhetoric, but all it was going to do was make people think we were a bunch of crazies," Lessig recalled.[24] Their partnership was short-lived, and Lessig sought out a more stable collaborator.

He found one in Eric Eldred. "You basically couldn't ask central casting for a better face," Lessig's colleague Jonathan Zittrain remarked of Eldred to the *Chronicle of Higher Education*. "There's none of the sort of ego that one stereotypically encounters with people on a mission."[25] In Eldred, Lessig saw a sympathetic defendant—not a moody loner, but a family man and lover of classic literature. "I try not to get involved in politics. I just like books and can't imagine not being able to share them with other people," Eldred wrote.[26] After the two men met for coffee and conversation, Eldred agreed to keep his site alive and sign on to the case. They filed suit in January 1999.

In October 2002, *Eldred v. Ashcroft*—John Ashcroft was, at the time, the US attorney general—was heard by the Supreme Court, prompting a rush of excitement among America's public-domain enthusiasts: a small but avid group of law professors, librarians, computer geniuses, and other bookish, persnickety souls. Like sports fans traveling to see their team compete for a title, many of them made the trek to Washington, DC, to support Lessig as he prepared to argue Eldred's case in the country's most important judicial venue. And what is the Rose Bowl, really, without a proper tailgate?

On October 8, the night before oral arguments began, a few of the *Eldred* pilgrims camped on the steps of the Supreme Court to ensure admittance the next morning. They approached the evening with a spirit of adventure and bonhomie: they played board games, sang songs, even ordered pizza. (Police officers helpfully provided the name of a pizzeria that would deliver to the courthouse steps.)[27] One of the campers, the Creative Commons cofounder Lisa Rein, brought a video camera and taped her companions as they explained what they were doing and why they had come to the court.

"Why did you fly out here from Chicago and come all this way to see the *Eldred* argument?" Rein asked one of her cohorts, a teenager in a brown jacket, who tugged on his chin in response, seemingly perplexed by the question. "It's very exciting to see the Supreme Court, especially such a prestigious case as this one," Aaron Swartz finally replied after he stopped fidgeting. Then he paused and tilted his head. "And . . . Larry invited me. Couldn't turn it down."[28]

AT the time, Aaron Swartz was almost certainly America's youngest public-domain enthusiast. A small and thoughtful fifteen-year-old from the wealthy Chicago suburb of Highland Park, Illinois, Swartz had flown to DC to watch oral arguments as a guest of Lawrence Lessig—with an assist from Lisa Rein, who had convinced Swartz's mother to let him attend. "When Lessig asked me if I was free that day to come, I laughed because I couldn't (and still can't) think of anything I'd rather do than attend," Swartz wrote at the time.[29] Though he already had a ticket for the oral arguments, he chose to spend much of the night shivering outside the Supreme Court alongside some of the only other people in the world who shared his special passion for open culture.

Aaron Swartz had always been an outlier. Even before he was acclaimed as the unofficial protégé of the copyright reform movement, Swartz was separated from his peers by the depth and breadth of his interests and intelligence, and the vehemence with which he expressed his unusual opinions. At school, he responded to roll call not with "Present," but with "I think, therefore I'm here."[30] He touted the virtues of *The Teenage Liberation Handbook*. His business card—he was the sort of boy who had a business card—billed him as "Writer, hacker, kid," and his line of work as "Emergent crypto anarchy."[31]

Swartz was born on November 8, 1986, and grew up with two younger brothers in what was surely the most computer-friendly household in Highland Park. His father, Robert, worked in the computer industry, and the Swartz home was filled with useful machines. "We were one of the first users of Netscape," Robert Swartz remembered. "We had an ISDN line, we were absorbed in this kind of technology when there weren't a whole lot of people doing it."[32]

Early on, young Aaron developed a keen interest in computers and information, and how the former could be used to acquire and organize the latter. ("I don't think I have any particular technical skills; I just got a really large head start," he later reminisced.)[33] Before reaching puberty, Swartz was writing simple computer programs and creating websites: for himself, for his family, for a local *Star Wars* fan club called Chicago Force. In 2000, when he was thirteen, Swartz built a website called The Info Network, which was a crowdsourced encyclopedia to which anyone could contribute. It was essentially the same concept as Wikipedia—but it predated Wikipedia by several months.[34] Swartz submitted The Info Network for the ArsDigita Prize—a youth Web-design competition sponsored by the entrepreneur and MIT researcher Philip Greenspun—and was named a finalist. He traveled to Cambridge, Massachusetts, to claim his prize.

That same year, the *Chicago Tribune* ran a story about the young prodigy and his online encyclopedia. "Getting 'real information' to people on the World Wide Web is 13-year-old Aaron Swartz's job. He's tired of all the banner ads, the sponsorships and other miscellaneous 'junk' hogging the screens," the article began.[35] "That's not what the Internet was made for. It was based on open standards and freedom," Swartz explained to the reporter. Swartz gravitated to others who felt the same way.

Around the same time he founded The Info Network, he also began contributing to online message boards and mailing lists for people who wanted to make the Internet more functional, which in turn, they hoped, would make the world a better place. "We were part of an inchoate, ad-hoc community of collaborators who helped each other learn how to code. No, not how to write code—how to write code *for the purpose of changing the world*," Swartz's friend Zooko Wilcox-O'Hearn later remembered.[36] Swartz barely paused to test the water before jumping in.

Wilcox-O'Hearn was already an adult when he made Swartz's virtual acquaintance. So were all the other people who haunted these lists, discussing open information and Internet usability issues. Most of them were professional programmers and academics who worked on these problems for a living. "You meet these people in text originally," recalled Dan Connolly, a software engineer and mailing list participant. "The guy's writing code, making intelligent comments; as far as you know, he's your peer. Then you find out he's fourteen, and you're, like, 'Oh!' "[37] These groups were, in a sense, modern-day inheritors of the AI Lab's hacker ethic. A participant's status was measured not by age or title, but by the quality of his contributions to the group.

Aaron Swartz didn't advertise his youth, and by the time that most of Swartz's correspondents discovered his juvenescence, they were already impressed enough by his intelligence that it didn't make much of a difference. "On the topic of not necessarily having a good feel for the age of net-based collaborators, I was blown away to learn that Aaron Swartz is in 8th grade :-!!," a developer named Gabe Beged-Dov wrote to an online mailing list on July 3, 2000.[38]

Swartz responded: "I generally try not to mention my age, because I find that unfortunately some people immediately discredit me because of it. :-(, Thanks to everyone who is able to put aside their prejudices not only in age, but in all matters, so that work on standards like these can go ahead and we can build the Web of the future. I don't know about all of you, but I get very excited when I think about the possibilities for the Semantic Web. The sooner we get standards, the better. It's not hard—even an 8th grader can do it! :-) So let's get moving."[39]

Swartz attended a private school, North Shore Country Day, in Winnetka, Illinois, and he chafed at its rules and customs. After-school sports were mandatory, much to his dismay. ("I narrowly escaped another day of practice due to an awful migraine headache. I don't know which is worse: the headache or practice," he blogged in August 2000.)[40] Students were burdened with too much homework and too many course requirements. Not only did his school inhibit intellectual curiosity, Swartz argued, it also seemed designed to stifle it. Also, the school's stairs were steep, and his backpack always seemed too heavy.

As he entered ninth grade in August 2000, Swartz launched a blog called *Schoolyard Subversion*, in which he portrayed organized schooling as a dystopia and himself as the leader of a burgeoning underage rebellion. "They drum it into your head. *Stop fighting now. You can't escape the message. Slowly, your brain shuts down. You stop thinking, stop challenging, stop asking questions," Swartz wrote with great melodrama. "You can't do it. You can't give in. You can't let them control you. You have to fight it, fight it every minute."[41]

Disenchanted students have long been fond of comparing high school to prison, but few have ever done so with such flair. While comparing an expensive private school to some sort of Orwellian dictatorship indicates a profound unfamiliarity with real institutional repression, Swartz was barely a teenager, and an unacculturated one at that. Though Swartz wasn't a social outcast, he never excelled at making close friends his own age. "I have developed my most meaningful relationships online," the fourteen-year-old Swartz wrote in April 2001. "None of them live within driving distance. None of them are about my own age. Even among those who I would not count as 'friends,' I have met many people online who have simply commented on my work or are interested by what I do. Through the Internet, I've developed a strong social network—something I could never do if I had to keep my choice of peers within school grounds."[42]

But most adults were not prepared to accept the young Swartz as a peer. Swartz's former principal, Robert Ryshke, remembers him as an assertive child who thought nothing of scheduling one-on-one appointments in which he would direct the bemused administrator toward books and articles about education reform. Even so, Swartz soon saw that the reforms he proposed were not imminent. High school students, he realized, were at the bottom of a fixed hierarchy, prevented from defining the contours of their own education. Every student, no matter how precocious, had to follow the standard path.

Swartz wanted to map his own route, and his parents allowed him to do so. "High school was for me probably the most unpleasant experience I ever had in my life," Robert Swartz would later say.[43] "So I was very sympathetic to the notion of Aaron's not wanting to go to high school." In the summer of 2001, Aaron gave notice: he would not return for tenth grade.

For the rest of his high school career, Swartz took classes at a local college and spent most of his spare time on the Internet, redoubling his involvement in online communities, where neither his age nor his maladroit manners held him back. Many of those in his online social network were involved with the World Wide Web Consortium, or W3C, a nonprofit group established by Tim Berners-Lee to promote technical standards and keep the Web running smoothly. Ten years after its public debut, the Web indeed resembled an infinite library, albeit a terribly disorganized one. The W3C members were among the first to understand the potential value and power of metadata as a solution to the search-and-retrieval problems that plagued the Web.

Just as a supermarket checkout machine scans a bar code to determine exactly what you're buying and how much it costs, a computer reads a website's metadata to acquire salient information about that site and the content therein. In a 2001 *Scientific American* article, Berners-Lee, James Hendler, and Ora Lassila made the case for a metadata-rich "Semantic Web" as one "in which information is given well-defined meaning, better enabling computers and people to work in cooperation. . . . In the near future, these developments will usher in significant new functionality as machines become much better able to process and 'understand' the data that they merely display at present."[44]

The idea sounded great to Swartz. "The future will be made of thousands of small pieces—computers, protocols, programming languages and people—all working together," he wrote in January 2002. "We need to stop worrying about how to make everyone do the same thing, and instead work on how to connect the (not so) different things that people do together."[45]

Swartz preached the virtues of Berners-Lee's project and became known in online communities for his youth, his intelligence, and his demeanor. "He seemed so confident about things even when he wasn't an expert," remembered Wes Felter, a software engineer who also participated in these groups.[46] "You couldn't just tell him, 'Well, you know, you're not an expert, you don't know what you're talking about. You're just a kid, you don't have experience.' He wouldn't take those bland arguments. He'd fight back."

Swartz's profile only rose once his correspondents started to meet him in person. His emancipation from high school had freed up his daily schedule, which allowed him to start occasionally attending computer conferences around the country. In the fall of 2001, accompanied by his mother, Susan, Swartz traveled to Washington, DC, for a conference sponsored by O'Reilly Media, a publisher of technical books and manuals. Susan Swartz brought Aaron to the conference and left him there in the care of Zooko Wilcox-O'Hearn, who had promised her that he would watch over her son. "I didn't sweat the 'keeping an eye on him' part much, but we did mostly hang around with each other that day," wrote Wilcox-O'Hearn, who was meeting Swartz for the first time.[47] "I remember being annoyed that he wouldn't eat his hamburger at lunch and he also wouldn't tell me why not."

Also at the conference was Lawrence Lessig, who, at the time, was almost three years into the *Eldred* case and was on the verge of opening a new front in the copyright wars. Lessig's new initiative had started when Eric Eldred suggested that somebody should establish a "copyright conservancy," a resource that collected material that had willingly been donated to the public domain. "If such an organization were set up, something on the order of The Nature Conservancy, then it would be something that authors and heirs could donate to themselves, when they have copyright to a work that is no longer published, but which deserves to be read still," Eldred wrote in 1998.[48] The idea eventually evolved into the distributed copyright conservancy known as Creative Commons.

Creative Commons was an effort to establish a more flexible framework for rights management in the digital era. One of the problems with copyright, Lessig reasoned, was that it was absolute. Once copyright was conferred, the rightsholder reserved *all* the rights to his creation. Creative Commons provided alternatives for people who perhaps wanted to reserve *some* rights to their creations, but were willing to cede others to the public.

The solution was a series of "licenses" that allowed creators to grant preemptive permission to people who wished to use and distribute their work. A photographer could publish one of her photographs under a Creative Commons "attribution" license, for example, which meant that anyone was free to copy, distribute, and display the photograph as long as they credited the photographer for her work. A "noncommercial" license bestowed usage and distribution rights as long as the photograph was not being used for commercial purposes. Creative Commons was not an attempt to replace statutory copyright—licensors retained the copyright to their works, and the licenses were fully compatible with existing copyright laws. But it was an unmistakable commentary on how statutory copyright often failed to meet the needs of the networked society. Copyright law needed to catch up with the times, Creative Commons implied—and if the laws lagged behind, then people would simply find a way to work around them.

When Lessig assembled a team to build the Commons, one of the first people he approached was the technology writer and archivist Lisa Rein. Rein, who was familiar with Swartz from his participation on various online mailing lists, suggested that Lessig ask Swartz to help implement the website's metadata. "I won't lie and say it wasn't hard at the time to convince these people that I needed this fourteen-year-old on the project," Rein said. "I took a pretty big hit for it politically at the time. Until they met him. It didn't make sense to anybody until they met him."[49]

When they met him, they were inevitably impressed—and occasionally they were intimidated. "He had incredibly high standards and we were not meeting them," remembered Ben Adida, a computer programmer who worked as a contractor on the project. "He was very critical of my work, and he said so publicly. The guy was pretty hard on me, and I didn't take that very well. He had a lot more of an audience on the Internet than I did, and it was hard for me to work with such a brutally honest person."[50]

Swartz was forthright about the ethicality of the Creative Commons project, too. He insisted that withholding access to information wasn't just bad policy: it was also morally wrong. "Without saying that what everybody else is doing is bad, we just thought these licenses might be good to help people share their stuff more," Rein remarked years later. "Aaron was ready to say, 'This system is bad, we need to change this.' "[51]

Swartz's dogmatism was sincere, but must have been derived, at least partially, from his environment. In less than a year, he had gone from complaining about the burden of a heavy book bag to collaborating with the intellectual elite of the open Web. Adults— important, smart adults—took his suggestions seriously. Swartz had joined a movement, and had even become a leader in that movement. These changes must have been exhilarating, and people moved by the fervor of exhilaration tend to yell the loudest.

In May 2002, Swartz traveled to San Francisco to attend another O'Reilly Media event, the first annual Emerging Technologies Conference, where Lessig and his cofounders planned to introduce the Creative Commons concept to a crowd that included some of the Internet's loudest yellers. For the fifteen-year-old Swartz, the trip was a succession of pinch-me moments. He spent much of his time in the hotel lobby, blogging and socializing alongside Wes Felter and other online friends. He joined them one evening at the Jing Jing restaurant in Palo Alto for a "Spicy Noodles Festival" organized by the software developer Dave Winer and the blogger Robert Scoble. (Swartz woke the next morning "with a neckcramp, a nosebleed, and a stomach-ache.")[52] He toured Google headquarters one afternoon at the invitation of employees there who admired his blog. "I had a wonderful time there," he wrote. "I got to meet lots of great Googlers and see all sorts of famous stuff."[53]

On the antepenultimate night of the conference, Swartz attended a party thrown by the writer Danny O'Brien and his housemates Jon Gilbert and Quinn Norton. There, he mingled with Bram Cohen, who designed the peer-to-peer file-sharing program BitTorrent; Cory Doctorow, who ran the massively popular blog *Boing Boing*; Jason Kottke, whose blog *kottke.org* was also widely read; and plenty of others. Though he was a fifteen-year-old boy at a party thrown and attended by adults, Swartz, for the most part, fit right in. ("It wasn't a really sophisticated scene, honestly," Wes Felter said of the era's tech-conference circuit. "[Swartz] was obviously less mature than other people there, but I would say not by a wide margin.")[54] At 11:00 p.m., Norton gathered everyone together to attend the midnight showing of George Lucas's new *Star Wars* film, *Attack of the Clones*. Swartz sat in the very first row. He was a long way from Highland Park.

On the final day of the conference, Lessig gave a speech drawn from the premises of his book *The Future of Ideas*. "Content providers launched a war to protect last century's way of business," Lessig said, in Swartz's paraphrase. "So far they've succeeded in stopping innovation. They've convinced the world that it's a choice between property and the American way or anarchy and evil communism. They're winning because that choice is simple. But there's something more."[55]

Something more referred to the Internet, and its unique decentralized architecture. "Technologists need to tell politicians the importance of the values held by the technology," continued Lessig, in Swartz's rendering of the speech. "They have a Valenti-like view of the world—'their terrorist war against the most important industry in America.' Rhetoric is right but the target is wrong. Technology is the most important industry. Need to let the folks back east understand this."[56] The remarks were a recruiting pitch, soliciting soldiers to join the cause of free culture. How could Aaron Swartz decline?

DESPITE Swartz's bravado, Lessig's words, and their colleagues' enthusiasm, "the folks back east" in Congress remained suspicious of file sharing, the Internet, and the public domain. "Not a single congressman will speak out against the interests of powerful computer or publishing companies," Eric Eldred predicted in 1998, and their reluctance was easy to understand.[57] From a politician's perspective, "the Internet will bring us intangible wealth and make us rich in knowledge" is a much less compelling argument than "the Internet is costing my campaign donors money."

The CTEA, the DMCA, and the NET Act were legislative attempts to thwart online copyright infringement. But in the years since their passage, online copyright infringement had ballooned into an international craze, largely thanks to the rise of peer-to-peer file-sharing services such as Napster, which allowed users to download digital music files for free from one another's computers, and which became ubiquitous in college dormitories soon after it launched. The Recording Industry Association of America (RIAA), the primary trade organization for record companies, claimed in 2002 that American manufacturers' earnings from the sale of recorded music had dropped by almost $2 billion since 1999, when Napster was released.[58]

In a July 2000 Senate Judiciary Committee hearing on the merits and demerits of online music downloading, Metallica drummer Lars Ulrich portrayed Napster as a digital den of unrepentant piracy. "Every time a Napster enthusiast downloads a song, it takes money from the pockets of all these members of the creative community," Ulrich said, remarking that the "touted new paradigm that the Internet gurus tell us we must adopt sounds to me like good old-fashioned trafficking in stolen goods."[59] Ulrich's position was clear: file sharing victimized songwriters, musicians, and record companies and threatened the survival of the entire music industry.

In *Moral Panics and the Copyright Wars*, the copyright scholar William Patry astutely noted that the mainstream culture industries operate on a mildly coercive "push marketing" model in which companies use advertising and promotions to create consumer demand for the products they want to sell, and the formats in which they want to sell them. Online file sharing repudiates "push marketing" by allowing consumers to unilaterally decide what they want to consume and how they want to do so. As file sharing grew ever more popular in the early 2000s, bringing with it potential opportunities for new, collaborative models of marketing and production, the culture industries instead focused almost wholly on ways to regain their lost control.

At that same hearing in July 2000, Gene Kan, one of the developers of the peer-to-peer file-sharing network Gnutella, offered a rebuttal, and a challenge, to those who claimed that the Internet was killing the music business: "Can we stem the tide of new technologies?" Kan asked. "Highly unlikely. So what does the future hold? Great things if profiteers adapt, if intellectual property profiteers adapt. Technology moves forward and leaves the stragglers behind. The adapters always win and the stalwarts always lose."[60]

As the *Eldred* oral arguments approached, the adapters and the stalwarts made their respective cases. The RIAA filed an amicus curiae brief that argued in favor of elongated copyright terms: "Petitioners' rhetoric masks what they are asking for: the right to copy—jot for jot—the works of others, often in order to profit economically," the trade group claimed. "Copyright piracy in the music business alone exceeds $4 billion annually. . . . Congress needs the flexibility to meet this threat effectively."[61]

The copyright conservatives didn't always frame their case in reactionary terms. They occasionally emphasized the social and cultural benefits of extended copyright, and all the good things that might be lost if Eldred and Lessig were to prevail. In another amicus brief filed in support of the federal government's position on *Eldred*, Dr. Seuss Enterprises argued that the long copyright terms provided by the CTEA promoted progress and creative expression. Without them, Americans would never have known the pleasures of Seuss Landing, a $100 million children's theme park built at Universal Studios Orlando: "At Seuss Landing, children can enjoy a 'Caro-Seuss-el' merry-go-round and other rides derived from such famous Seuss books as *The Cat in the Hat* and *One Fish Two Fish Red Fish Blue Fish*. This park would never have been created had the Seuss works not been protected by copyrights."[62] The implication was clear: assigning books into the public domain would deprive the world of theme parks inspired by those books.

No matter your opinion of the culture industries, it's true that the people employed in them tend to lose their jobs when revenues decline. There are valid economic arguments in favor of long copyright terms. But the entities making those arguments often seemed to willfully ignore any public benefit that might be derived from the public domain. In 2002, Brewster Kahle devised a novel way of illustrating that public benefit.

Kahle, an engineer and entrepreneur who had worked at the MIT AI Lab during the tail end of the hacker-ethic period, had been waiting for twenty years for technology and circumstances to catch up with his dream of building a computerized library that could rival the ancient Library of Alexandria in breadth and ambition. Kahle was gangly and bespectacled and looked like someone who might be hired to host a children's television program about science. In 1999, he sold his company, Alexa Internet—an homage to his beloved Library of Alexandria—to Amazon for $250 million in stock, and then turned his attentions to building and maintaining the Internet Archive, which he founded in 1996. The nonprofit Internet Archive is dedicated to the overwhelming task of archiving the entire World Wide Web. It sends little "spiders" spinning across the Web to "crawl" through every website they can find and to memorize what those sites looked like on any given day. Those snapshots are then stored on the Internet Archive's servers, where they serve as a massive, functional photo album of the World Wide Web past and present.

For Kahle, however, archiving websites for posterity had always been a prelude to the archiving of books. In late summer 2002, Kahle began uploading public-domain books onto the Internet Archive servers. Then he purchased an old Ford minivan and christened it the Internet Bookmobile. On the side of the bookmobile, written in the Comic Sans typeface, was the phrase 1,000,000 Books Inside (soon). Inside the bookmobile were a couple of laptop computers, a high-speed color printer, and a bookbinding machine; on its roof sat a satellite dish connected to the Internet Archive's servers in California. That fall, Kahle packed his eight-year-old son, a couple of friends, and a freelance journalist named Richard Koman into the bookmobile, and drove it cross-country in a mobile demonstration of the good things that can happen when the public domain meets an eccentric, civic-minded multimillionaire.

Kahle made stops in Salt Lake City, Columbus (Ohio), Akron, Pittsburgh, and Baltimore. He also stopped in Urbana, Illinois, to surprise Michael Hart, but Hart was loath to leave his house. ("When we arrive at his house, there is a car parked in the driveway but no other signs of life," Koman wrote. "A sign on the front door says 'RING BELL LOUD. RING AGAIN. PAUSE. THEN RING AGAIN.' Following these directions yields no response.")[63]

At every stop, Kahle and his bookmobile gathered a crowd. "Woohoo! We're making books!" he cried, and that's exactly what he proceeded to do. A user could connect to the Internet Archive's servers, download a public-domain text, print it out, and run it through the bookbinding machine. After fifteen minutes, the user owned a brand-new book. "The books aren't perfect: There are a few typos, some bad line breaks, and straight quotes instead of curly quotes, but they still look remarkably good," Koman reported. They were inexpensive to produce, too: "In a print-on-demand world, where the cost of creating a book runs about $1 and the capital costs run under $10K, libraries don't lend books, they give them away."

Kahle's road trip concluded in Washington, DC, right before Lawrence Lessig delivered the *Eldred v. Ashcroft* oral arguments. On his own blog, Lessig hailed Kahle's vehicular agitprop, noting "the difference between a brilliant mind and a lawyer's mind. While we brought a law suit [*sic*], Brewster built a bookmobile."[64] Though he had spoken in passionate tones at the Emerging Technologies Conference, Lessig was much less strident when bringing *Eldred v. Ashcroft* to the Supreme Court, instead seeking to "frame the *Eldred* case in a very sterile constitutional way."[65] In his opening brief, Lessig had argued that, by passing the Copyright Term Extension Act, Congress exceeded its constitutional mandate. The framers had intended copyright to be a limited-time right, and the retroactive nature of the CTEA violated that provision. As he prepared to address the nine justices, Lessig tried to maintain this professorial dispassion. *Eldred* was "the rare case where the law, properly and carefully read, yields one right answer," and Lessig was determined to lead the court to that answer through reason rather than rhetorical appeal.[66]

His colleagues were more than happy to pick up his declamatory slack. In the months preceding the *Eldred v. Ashcroft* arguments, Swartz's blog posts continued to frame the copyright question as a moral matter: "I've often complained to folks about their use of the term 'pirate' to mean 'share,' " he wrote in February 2002. "When folks complain about pirated movies, do they really mean to imply that sharing movies with someone is the moral equivalent of attacking a ship?"[67] He bragged that he was using a service called LimeWire to download copyrighted music for free.[68] On his way to Washington for the oral arguments, Swartz fantasized about encountering Jack Valenti in O'Hare Airport and asking the copyright lobbyist the sorts of hard questions that would leave him "mumbling and looking down at his watch."[69] The real debate, on the courtroom floor, would not so easily be won.

The day before oral arguments were scheduled to begin, Lessig delivered a pep talk to a group of supporters. A few days later, Swartz paraphrased those remarks on his blog:

> Four years ago, when we filed this case, people laughed us out of their office. "You want to take away people's property?" they exclaimed. No one understood what the public domain was, the media thought we wanted to get rid of copyright. That's not the case now. Every article understands the issues, people know what the public domain is. That's an important victory.
>
> Even more important is that we have a group like this. We've got a team of people here fighting for our freedom. Whatever happens tomorrow, whatever the court decides, let's not lose this, let's not stop the momentum. There are many battles to fight, and we need to keep going.[70]

Later that night, Swartz returned to the Supreme Court steps, where he talked and laughed and played the board game SET with the other members of his tribe, waiting for morning to come and the world to set itself right.

The next morning, the courtroom was completely full. The Federal Reserve chairman Alan Greenspan was there; so was *Hackers* author Steven Levy. Jack Valenti was there, as was Sonny Bono's widow, Representative Mary Bono. "The courtroom itself was an impressive structure," Swartz noted on his blog. "Everything was very, very tall."[71]

The court seemed skeptical of Lessig's argument. "Many Justices repeatedly said that they felt [the CTEA] was a dumb law, that it took things out of the public domain without justification," Swartz reported later on his blog. "But they were having trouble finding a way to declare it unconstitutional without also having to overturn the '76 extension, something they clearly didn't want to do."

Lessig had argued a case in open court only once before, and his courtroom inexperience was evident. He was too much the professor, and not enough the performer, it seemed to Swartz. "I thought Larry had done an awful job until Solicitor General Olson (the man who argued for Bush in *Bush v. Gore*) came up," Swartz wrote. "The Justices had a field day with him. Rehnquist got him to admit that a perpetual copyright would violate the Constitution. Kennedy got him to admit that a functionally perpetual (900 year) copyright would also be a violation." Neither Olson nor Lessig had clearly triumphed, and as oral arguments concluded, the case seemed to be either side's to win. Eldred later summarized his day in court for the members of the Book People mailing list. "Our side was ably argued and the government's case was very weak," he wrote, noting that "the calls from reporters are petering out, so maybe now I can go back to scanning books."[72]

That afternoon, Swartz and his friend Seth Schoen visited the

Library of Congress and toured the Main Reading Room. Similar to the Boston Public Library in 1895, the ornate and sumptuous gallery seemed to apotheosize the notion that unfettered information was the guarantor of American liberty. "Seth exclaimed that the LOC Reading Room was the most beautiful thing he'd ever seen," Swartz reported, but the setting made both of them emotional.[73] Hours earlier, Lessig had issued the Supreme Court a challenge to greatness. Standing inside the Reading Room, acolytes in a temple to the transformative power of the written word, they could hardly doubt that the court would rise to meet it.

ON January 15, 2003, the Supreme Court ruled 7–2 in favor of Ashcroft. In her majority opinion, Justice Ruth Bader Ginsburg wrote that the Sonny Bono Copyright Term Extension Act was "a rational enactment" and that the court was "not at liberty to second-guess congressional determinations and policy judgments of this order, however debatable or arguably unwise they may be." Lessig's contention that a consistently extended copyright term effectively constituted perpetual copyright was unpersuasive, Ginsburg opined. "Beneath the facade of their inventive constitutional interpretation, petitioners forcefully urge that Congress pursued very bad policy in prescribing the CTEA's long terms," she concluded. "The wisdom of Congress' action, however, is not within our province to second-guess."[74]

That was that, and as the industrial copyright stakeholders celebrated their victory, their opponents struggled to process the predictably unhappy outcome. "It's Over. We Lose," Lisa Rein headlined her blog post on the *Eldred* decision. "So the Public loses again. Par for the course these days."[75] On the website *Boing Boing*, the writer Cory Doctorow expressed his sorrow: "This blog will be wearing a black armband for the next day in mourning for our shared cultural heritage as the Library of Alexandria burns anew."[76]

"We'd be interested in your ideas about what to do next," Eric Eldred wrote to the Book People mailing list hours after the court released its decision.[77] While some respondents proffered new strategies and tactics, others just seemed frustrated and sad. "Obviously the goal is to have virtually no public domain left at all," Michael Hart wrote.[78] Lessig's gambit had failed, and the public-domain enthusiasts could reasonably wonder whether they ever had any chance of winning a game that seemed rigged in favor of their opponents.

Lessig later acknowledged that he had mistakenly approached the *Eldred* case as a strict legal question, instead of framing it as a moral issue. "I have stood before hundreds of audiences trying to persuade; I have used passion in that effort to persuade; but I refused to stand before [the Supreme Court] and try to persuade with the passion I had used elsewhere," he admitted in 2004's *Free Culture*, a book-length attempt to reargue the *Eldred* case and extrapolate its principles across other modes of culture.[79] Dispassion might be the standard in academic circles, but political outcomes are rarely decided by strict appeals to reason. American copyright law is a product of morality and metaphor. It is difficult to win a game that you choose not to play.

Aaron Swartz's response to the Supreme Court's decision was uncharacteristically subdued. On his blog, he linked to some coverage of the verdict, then added a single optimistic line of his own: "If we cannot overturn it in the courts, then we shall overturn it in the legislatures."[80] In January 2003, this outcome did not seem particularly likely. The copyright industries had all the advantages on their side: lots of money, a gaggle of experienced lobbyists, and a century's worth of legislative precedent.

The free culture advocates had energy and time. If nothing else, *Eldred v. Ashcroft* was a chance for them to declare their existence; to announce their cause to the federal government and certain spirited segments of the American public. "We're an emerging society!" Lisa Rein exclaimed as the free-culture group set up camp outside the Supreme Court the night before oral arguments began.[81] She was right. This society was emerging from the depths of the Internet. And Aaron Swartz was emerging as one of its foremost citizens.

6

"CO-OPT OR DESTROY"

During his "unschooling" years, uninhibited by other people's standards, Aaron Swartz pursued his own interests and developed his own identity, more so than most teenagers of his generation. "Unschooling has been great for me. I've never felt so relaxed or at peace with myself before," Swartz wrote in December 2001. "While making it work and finding things to do have been difficult, I've been forced to sort out my priorities and figure out how I work best. I doubt this would have happened in school, where you are told what's important and when it is due."[1]

But if withdrawing from high school because it was insufficiently challenging made you a prodigy in the eyes of the world, forsaking college entirely just made you look lazy. In the fall of 2003, during what would have been his senior year at North Shore Country Day, Swartz prepared to reenter the world of mainstream education. Stanford University, in Palo Alto, California, seemed like the most promising option. Lawrence Lessig had relocated there. It sat at the heart of Silicon Valley and was close to San Francisco and all of Swartz's friends in that city's free culture scene. So Swartz applied. Tim Berners-Lee wrote him a letter of recommendation. On the day Swartz received his acceptance letter, he noted the occasion on his blog with exaggerated glee: "so, like, i totally asked Stanford out, like a whole month ago, and today she said YES! omg!!!!!!! so cool!!!"[2]

This mock enthusiasm soon subsided. On his first night at Stanford, orientation leaders schooled the new arrivals in "the dos and don'ts of campus life. You know: smoke your pot over by the lake, keep your vomit from binge drinking off the floor, and never, ever share files over the Internet. (To underscore the last point, we were given an MPAA flier along with the key to our room.)"[3] By his third day in Palo Alto, Swartz's tone already indicated his fear that he had made a mistake. "It's hard to say this without sounding even more superior than usual, but it doesn't strike me that most Stanford students (and professors) are exceptionally bright," Swartz posted on his blog. "I was led to believe that Stanford was a magical place where everyone was a genius. This is somewhat disappointing."[4]

As the months passed, Swartz's disappointment mounted. His classes were boring, his fellow students insufficiently serious. He made few friends. Stanford was North Shore Country Day with more people and better weather. As a teenager in Illinois, Swartz had valued his online friendships in part because so few of his local peers shared his interests. "I remember he was really hoping that that'd change when he went to Stanford," his friend Seth Schoen recalled. "And I visited him there and he basically said that it hadn't—that even the other Stanford undergrads around him weren't curious about the things he was curious about."[5]

By Swartz's own account, his freshman year was a social disaster, an adventure in loneliness and miscommunication. On his blog, he chronicled college life, adopting the tone of a priggish anthropologist studying some vulgar foreign culture. At times, his isolation is painfully evident:

Stanford: Day 58

Kat and Vicky want to know why I eat breakfast alone reading a book, instead of talking to them. I explain to them that however nice and interesting they are, the book is written by an intelligent expert and filled with novel facts. They explain to me that not sitting with someone you know is a major social faux pas and not having a need to talk to people is just downright abnormal.

I patiently suggest that perhaps it is they who are abnormal. After all, I can talk to people if I like but they are unable to be alone. They patiently suggest that I am being offensive and best watch myself if I don't want to alienate the few remaining people who still talk to me.[6]

Swartz may not be the most reliable narrator regarding his freshman year. His former roommate does not remember him as a social pariah. Another friend, Zooko Wilcox-O'Hearn, described on his own blog a since-deleted posting Swartz wrote about necking with a random girl one night on the Stanford campus. "I can't believe I'm dry-humping Aaron Swartz!" the girl said, in Wilcox-O'Hearn's recollection. "I can't believe I'm dry-humping!" Swartz replied.[7]

Swartz may not have been quite as isolated as he claimed to be, though he clearly wasn't fulfilled. While his fellow students had ostensibly enrolled in college to find themselves, Swartz mostly found himself ever more separated from the things that made him happy. While back in Highland Park during winter break, Swartz attempted to articulate exactly why he found organized schooling so enervating. "When I started high school, I remember watching for that point where foreground and background reverse—the point at which school as a use of my time turned into time being what was left over after school. It didn't take long," he wrote on his blog. "School is like that. It keeps you running until running is the only thing you know."[8] Not surprisingly, he took the first possible opportunity to run away from Stanford.

Swartz was exactly the sort of disaffected, understimulated genius that Paul Graham was hoping to recruit for his latest enterprise. Graham is a computer programmer and entrepreneur who, in 1998, sold the company he cofounded, Viaweb, to Yahoo for $49.6 million in stock. Following the sale, Graham wrote a series of thoughtful essays on computers and the people who loved them. In 2003, Swartz had excerpted one of those essays, "Why Nerds Are Unpopular," in which Graham suggested that the pointless busywork assigned in the typical American high school only encourages smart, self-motivated teenagers to consider suicide. "Another problem, and possibly an even worse one," continued Graham, "was that we never had anything real to work on. Humans like to work; in most of the world, your work is your identity. And all the work we did was pointless, or seemed so to us at the time."[9] ("I highly recommend reading the whole thing," Swartz added.)

Graham also aspired to invest in start-up companies. In 2005, he thought he had figured out what caused so many tech start-ups to fail. A short five years after the American economy had sunk under the weight of a million irrelevant websites, the obvious answer to the question was "Because those websites were irrelevant." But *why* had they been so irrelevant? Why had the tech bubble been characterized by smarmy, overcapitalized mistakes, vague solutions to problems that no one had posed, and companies that wasted their venture capital funding on office whimsies and art direction while neglecting to ensure that their websites actually worked?

Well, for one thing, Graham observed, in that age of irrational exuberance, the websites didn't actually have to work in order to attract investors. But a bigger problem was that venture capitalists had invested in the wrong sorts of entrepreneurs. Investors were attracted to companies founded by people with significant business experience, people who would have been just as comfortable working in the auto industry as at theautoindustry.com. Graham suspected that "success in a startup depends mainly on how smart and energetic you are, and much less on how old you are or how much business experience you have."[10] He wondered what would happen if venture capitalists bypassed the businessmen and, instead, invested directly in the ideas of smart kids who could build stuff. "Animals," Graham called them: tenacious and intelligent young self-starters who didn't need Ping-Pong tables in the office as long as they had a case of Dr Pepper and a reliable Internet connection.

Graham decided to test his hypothesis and, in March 2005, announced that he was soliciting applications for a project he called the Summer Founders Program—an early version of what would eventually become the renowned start-up incubator Y Combinator. "The SFP is like a summer job, except that instead of salary we give you seed funding to start your own company with your friends," he wrote on his blog.[11] Aspiring young entrepreneurs proposed ideas for start-up companies to Graham; the most promising applicants would be invited to move to Cambridge, Massachusetts, and participate in a sort of start-up summer camp. The participants would each receive approximately $6,000 in seed funding and would spend the summer developing their businesses and learning from Graham and his well-connected friends. Come fall, the Summer Founders, theoretically, all would have developed prototypes and business plans and the acumen necessary to go out and seek serious start-up funding. "If that sounds more exciting than spending the summer working in a cube farm," Graham concluded, "I encourage you to apply."[12]

Swartz submitted a proposal for a service called Infogami, a tool for quickly building customizable, visually interesting, wiki-enabled websites.[13] (The name *Infogami*, which rhymes with *salami*, is a compound of the words *information* and *origami*.) "The Macintosh completely changed the way people used personal computers. Instead of typing arcane commands, you could point at what you wanted. Instead of just being able to work with text, anyone could use it to do graphics," Swartz would later write. "Infogami is the Macintosh of building websites. Perhaps it's a bit lofty of a goal, but we say aim high."[14]

Graham was intrigued by Swartz, if not necessarily Swartz's idea, and so he invited the restless Stanford undergrad to come to Cambridge and pitch the concept in person. As Swartz packed for the trip in his Stanford dorm room, he told his roommates that he was off to interview for a summer job. His friend Seth Schoen, amused at the understatement, suggested Swartz explain that the interview was with Paul Graham, the famous programmer and essayist. "Yeah," Swartz said, "but they won't know who that is."[15]

On his blog, Swartz portrayed the pitch meeting as a comical and bemusing experience. In Swartz's telling, Graham spent the session bouncing between conversational topics while paying surprisingly little attention to Swartz's proposal for Infogami—"which he appears not to have read very carefully," Swartz noted. Instead, the investor seemed primarily interested in convincing Swartz to abandon his original idea in favor of one of Graham's own—or, failing that, to at least change his company's name from Infogami to something more pronounceable. Nonetheless, Graham's hyperkinesis at least indicated an excess of curiosity and ambition, a welcome change from the lethargic Stanford scene. "I walk down the sunny Cambridge street, smiling," Swartz wrote after the pitch session concluded. "I feel pretty confident of being accepted."[16] His hunch was right. Graham called him back that same night and welcomed him to the Summer Founders Program. After a year on the academic treadmill, Swartz had regained a sense of forward motion.

He returned to California the next evening and arrived at his dorm room to find his roommates busy on their computers, unaware that Swartz's life had just changed. Swartz didn't bother to tell them. "I put my computer away and go to sleep," he wrote, "sleeping the sleep of a man who, whatever his surroundings, knows that at heart he is a capitalist."[17]

BUT Swartz proved to be no more adept at capitalism than he was at college. When Stanford's spring semester wrapped up, Swartz moved to Cambridge to begin work on Infogami alongside one collaborator, the young Danish programmer Simon Carstensen, whom Swartz had met online. "Aaron ended up renting this dorm room at MIT, and that's where we worked over the summer on Infogami," said Carstensen.[18] The room was small and hot, and the sense of sweaty claustrophobia was heightened because the two men were virtual strangers. "When I arrived in Boston, I knocked on the door [of the MIT dorm room]. We both looked at each other. We'd never met. And we were going to be here for two months," Carstensen recalled. "It was weird."

It remained weird. As the summer progressed, the eighteen-year-old Swartz's deficiencies as a manager and collaborator became evident. Swartz didn't trust Carstensen's code and rewrote much of it himself. "It was a very tough summer," Carstensen remembered, and when it was over he returned to Denmark. "[Swartz] asked me to keep working on the project, and I decided not to," said Carstensen. Unfazed, Swartz decided to stay in Cambridge and to keep building Infogami by himself. His inability to successfully do so made him frustrated and depressed. "The whole experience was incredibly trying. There were many days when I felt like my head was going to literally explode," he wrote. "One Sunday I decided I'd finally had enough of it [Infogami]. I went to talk to Paul Graham, the only person who had kept me going through these months. 'This is it,' I told him. 'If I don't get either funding, a partner, or an apartment by the end of this week, I'm giving up.' Paul did his best to talk me out of it and come up with solutions, but I still couldn't see any way out."[19]

Eventually, Graham suggested that Infogami merge with another understaffed Summer Founders start-up: a social bookmarking website called Reddit, which let users share and discover links to interesting online content. Graham had pitched Swartz to Reddit's two original founders, Alexis Ohanian and Steve Huffman, as a savvy programmer who could help them develop their website and take it to new heights. Swartz, meanwhile, could draw on the Reddit team for help with Infogami. Ohanian and Huffman bought in, and by November 2005, Infogami and Reddit had merged, forming a new umbrella company called Not a Bug. But what at the time felt to Swartz like a fresh start soon turned out to be yet another disappointment.

"There was a time, for a couple of months, when we were, like, 'Okay, we're gonna start this new thing, and it's gonna be bigger than Reddit, bigger than Infogami,'" Huffman recalled.[20] "And then it became pretty clear a few months in that that was not going to be the case." Ohanian and Huffman were recent graduates of the University of Virginia and close friends. Swartz was an introvert with an interloper complex. As he wrote on his blog in 2005, "I'm afraid of asking for things from people, even the tech support guy on the phone; I'm excellent at managing m[y] own free time, and thus distasteful of structured activities; I have trouble making friends with people my own age; and I hate competition."[21]

After an initial burst of productivity, during which Swartz and Huffman collaborated to rewrite Reddit's code in the Python programming language and build a backend database that could support both Reddit and Infogami, Swartz's working relationship with his new colleagues deteriorated. Infogami lay fallow as the more popular Reddit became Not a Bug's top priority, and Swartz was dispirited by this outcome. "It's not that I don't enjoy my work; it's just that I feel like I'm getting dumber doing it," he wrote.[22] While his colleagues spent their days and nights programming the website and bootstrapping the business, Swartz consciously played third wheel, working inconsistently on Reddit while pursuing other interests: running for a seat on the board of trustees of the Wikimedia Foundation, sitting in on random college lectures, incessantly blogging. He lost weight experimenting with fad diets. ("Friends and acquaintances urge me to eat more, doctors think I'm sick, family members suggest I have an eating disorder," he wrote.)[23] He acquired a sleeping bag and spent an evening in mock-homelessness, sleeping outdoors in the recessed entrance of a bookstore in Harvard Square.[24] "*What was so hard about that? I thought.*"

Swartz's privileged youth showed itself in moments like these.

"My grandfather was a capitalist. My father was a capitalist. I went to elementary school and junior high in the sixth-richest city in America. I went to high school in the third-richest," he wrote in April 2005.[25] His parents had had the wherewithal to underwrite his youthful exceptionalism; he had been free to opt out of systems that did not regard him as special. It is easy to sleep on the street when you know you are doing so by choice; it is easy to shirk tedious tasks when your well-being has never hinged on their completion.

Aaron Swartz was nineteen years old when Reddit and Infogami merged, ferociously intelligent but inexperienced with life, thrust into an environment that demanded more maturity and commitment than he could supply. Not a Bug needed him to be one thing: a programming animal. Swartz refused to accept those constraints. "I don't *want* to be a programmer," Swartz wrote on his blog in May 2006.[26] "When I look at programming books, I am more tempted to mock them than to read them. When I go to programmer conferences, I'd rather skip out and talk politics than programming. And writing code, although it can be enjoyable, is hardly something I want to spend my life doing."

Swartz harbored more intangible ambitions. As early as March 2005, back before he had left college to chase start-up riches, Swartz had described his burgeoning disillusionment with the programmer's life. When an acquaintance asked Swartz why he switched majors "from computer science to sociology, I said it was because Computer Science was hard and I wasn't really good at it, which really isn't true at all. The real reason is because I want to save the world."[27]

Since the days of Jean-Jacques Rousseau, young men and women have yearned to save the world: to live according to their ideals and spend their days fighting social injustice and effecting substantive change. This dream is usually abandoned with age, deferred into idle bar-stool radicalism and the occasional protest vote. The world has many problems, and even the most effective individual would be unlikely to solve all of them absent divine intervention or black magic.

The novelist Arthur C. Clarke's famous Third Law states, "Any sufficiently advanced technology is indistinguishable from magic." It is also true that any sufficiently advanced technology encourages magical thinking. *New* technologies are indistinguishable from magic wands, imbued with great and implausible powers that can transcend societal barriers and the laws of thermodynamics. Silicon Valley is rife with examples of aspiring messiahs touting the world-historical potential of their products. These promises are generally unfalsifiable, which is why they are simultaneously so attractive and so empty.

The informaticists Rob Kling and Roberta Lamb have argued that "talk about new technologies offers a new canvas in which to reshape social relationships so they better fit the speaker's imagination."[28] Since his early teens, Swartz had imagined a world free from "laws that restrict what bits I can put on my website,"[29] a world where culture could not be owned.[30] Occasionally—only just— Swartz encountered other people his age who felt the same way.

In June of 2005, just before relocating to Massachusetts, Swartz traveled to Washington, DC, to attend a conference on free culture. There, he met three young political activists from Worcester, Massachusetts—Holmes Wilson, Nicholas Reville, and Tiffiniy Cheng— who had been living out their ideals for a while and had enjoyed some success at grabbing the world's attention. Swartz had met Wilson before, in San Francisco, and had marked their meeting on his blog, writing that Wilson was a "very cool guy," but, surprisingly, "less radical than Downhill Battle's extremism led me to believe."[31]

The advocacy group Swartz referred to, Downhill Battle, argued vehemently in favor of digital file sharing. Wilson and Reville had founded it in 2003, just as the Recording Industry Association of America (RIAA), like a dying wasp reflexively stinging everything in its flight path, tried to frighten America's digital music downloaders away from their computers and back to Best Buy where they belonged. In that year, the RIAA filed suit against 261 Internet users nationwide whose online history indicated that they had illegally shared music files over the Internet; the trade organization sought exorbitant financial damages in reparation. (A brief personal aside: One of the people who was sued was my mother, whose name was on the Internet account that my teenage sister used to illicitly share some tracks by Destiny's Child and O-Town. My mother has never illicitly shared anything in her life. My mother was named Woman of the Year by her church.)

The Downhill Battle activists mounted a flamboyant protest of the RIAA's tactics. They designed some black-on-red stickers reading "WARNING! Buying this CD funds lawsuits against children and families"—a grassroots version of the famed "Parental Advisory" tags. Then they visited various chain retailers and surreptitiously affixed the agitprop to unsold CDs by major-label musicians. Their efforts didn't end with principled consumer vandalism, though: Downhill Battle also created a "Peer-to-Peer Legal Defense Fund" to help individual defendants contest the RIAA lawsuits.

The group found other targets for its ire, too. When the iTunes online music store launched in 2003, for instance, Downhill Battle released a parody of the iTunes interface dubbed "iTunes iSbogus" and argued that the Apple music marketplace exploited both artists and consumers. Rather than use the Internet to create a better music-sales system that returned a higher percentage of profits to singers and songwriters, iTunes simply duplicated the existing model, in which the bulk of the profits went to the distributors. This wasn't innovation, Downhill Battle argued; this was the same imbalanced system in a new, streamlined skin. "If you want to support the musicians you love, the best way to begin is by downloading the song for free on a filesharing network," they wrote. "Then send them what you want to give, no middleman."[32]

This particular idea was more than a little starry-eyed. "If you want to make digital music distribution work, then send your favorite musicians whatever you want, if you want, no obligation" isn't a business model so much as a magic wish. While a few musicians would, years later, find some success with the "pay what you can, and don't worry if you can't" strategy, most of these had already developed large fan bases during their years with major record labels. Emerging artists would find it exceedingly difficult to earn a sustainable living from such an ad hoc business model.

Nevertheless, the members of Downhill Battle were living out their principles, and Swartz admired them for it. After a full year at Stanford among young people who neither knew nor cared about the intellectual justifications for free culture, Swartz was invigorated by the Worcester group. "At a time when most copyright activists were painstakingly careful to insist they opposed downloading copyrighted music and arguing that the music industry should find a way to make downloading music legal, the website's opposition to one of the first moves in this direction and its full-throated support of file-sharing software was nothing less than shocking," Swartz wrote of Downhill Battle's position.[33] The young activists aggressively defended their beliefs and fearlessly engaged with their ideological opponents. They were comfortable with themselves and their place in the world.

Swartz could not say the same for himself. He got along reasonably well at tech conferences, surrounded by introverted computer programmers in GNU Project T-shirts and cargo shorts. But he was clumsy and self-conscious around most other people his own age. His cringe-worthy account of his first collegiate party—"Teenagers moving their bodies in bizarre and vaguely rhythmic positions in close proximity to one another. I'd seen the practice frequently enough on TV, so on one level I knew what to expect, but on another it was wholly bizarre"[34]—underscored his physical and social distance from his age-group peers.

He wrote of attending a house party with the Worcester people and retreating alone to the kitchen "with a neck ache" when the music swelled, the lights dimmed, and most of the attendees began to dance. Swartz was used to self-segregating while other people had fun; this was likely a way to minimize his social anxieties and camouflage his discomfort with his own body. At this particular party, however, Holmes Wilson came into the kitchen and asked Swartz if he was happy sitting on his own. "Because I want to know if I should pressure you to come join us and I know there are some people who are just totally OK with who they are," said Wilson.

Swartz paused. "Well, right now, this person I am right now, has no desire to go dancing. But I'm beginning to wonder if I should be a different sort of person."[35]

Over the weekend of the conference, Swartz and the Worcester activists explored Washington, DC, going to restaurants, skateboarding—"things go too fast and I arch my back and scrape my wrist and knee," Swartz wrote. They even sneaked the eighteen-year-old Swartz into a bar for the first time in his life. ("It's quite stylish, with globe lights and fake branches as plants," Swartz observed.) Swartz's account of the experience was one of the longest and happiest blog posts he had ever written, evincing a palpable sense of self-discovery. "I hang out with Tiffiniy on the front steps for a bit and then she grabs my hand and drags me across the street and over a fence, up some stairs into this amazing urban meadow between two houses that's been allowed to run wild, grass growing tall and crawling up a defunct fountain in the middle," he wrote. " 'Isn't this beautiful?' she asks and indeed it is. After some time we head back and some time after that she heads off with some people to a bar. 'Do you want to come?' she asks and I do but I also don't want to cause trouble again so I stay behind and head back to the hotel, looking at myself in the hotel's many mirrors, checking out how I look in the tight-fitting Downhill Battle tee they gave me."[36]

In a time of life when Swartz was trying on identities, "free culture activist" seemed to fit him well.

IN the middle of 2006, Condé Nast, the parent company of *Wired*, the *New Yorker*, and many other magazines, expressed interest in acquiring Not a Bug. Despite the tensions between Swartz and his colleagues, Reddit attracted roughly five hundred thousand unique visitors per month.[37] Although most of the site's users probably didn't realize it, they were Reddit's product. Their attention and loyalty could be sold to advertisers eager to promote their goods on the site; their browsing and sharing histories could conceivably be mined for a wealth of personal information that could also be sold off. Content producers could, theoretically, even pay to have Reddit show users their stories. If attention was currency on the nascent social Web, then Reddit seemed primed for riches.

The prospective acquisition presented Swartz with an existential quandary. Though he claimed to not care about money, he didn't object to having or making it. But he wondered whether he and his colleagues actually deserved the sums under discussion. On his blog, Swartz openly questioned Reddit's real value, recounting a conversation with an author who was astounded at the site's popularity:

> "So it's just a list of links?" he said. "And you don't even write them yourselves?" I nodded. "But there's nothing to it!" he insisted. "Why is it so popular?"
>
> Inside the bubble, nobody asks this inconvenient question. We just mumble things like "democratic news" or "social bookmarking" and everybody just assumes it all makes sense. But looking at this guy, I realized I had no actual justification. It was just a list of links. And we didn't even write them ourselves.[38]

Just a list of links. That assessment wasn't quite accurate. Years later, Reddit would develop into one of the Internet's most robust community platforms, a site that brings disparate users together to form interest groups, converse, and collaborate. In 2006, though, Swartz didn't foresee this massive popularity—or, if he did, he wasn't about to wait five years for the site to find its legs. A list of links wasn't what Swartz had left college to work on. "You can say a site is cool, stupid, popular, a flop, innovative, or clichéd," Swartz wrote on his blog. "But the one thing you can't say, the one thing that everybody skips over, is that these sites aren't anything serious."[39] Silicon Valley seemed, to Swartz, to operate on an inverted moral calculus that granted start-ups rewards disproportionate to the value they added to the world. Swartz was leery of acculturating to that sort of environment.

The summer before he enrolled at Stanford University, Swartz had read a book called *Moral Mazes*, an ethnographic study of middle management at several large corporations. (He later called *Moral Mazes* his favorite book of all time.) On his blog, Swartz remarked that the book "tells the story of a company, chosen essentially at random, and through careful investigation from top to bottom explains precisely how it operates, with the end result of explaining how so many well-intentioned people can end up committing so much evil."[40]

The author of *Moral Mazes*, the sociologist Robert Jackall, described the middle manager's world as one characterized by compromise and justifications; a system in which managers are often rewarded for doing things that are bad for society but good for their employer. Most institutions, he demonstrated, were built for self-perpetuation. Their employees and partisans were expected to act with the health of the institution in mind, regardless of what the consequences of their actions might mean for the general welfare. The central lesson of the book, according to Swartz: "Corporate managers simply aren't allowed to be moral, or even reasonable. And those who try are simply weeded out."

Swartz was thus ambivalent about the Condé Nast deal, about potentially becoming a cog in a corporate machine. The negotiations lasted for months, and they just exacerbated the tensions that had grown between Swartz, Huffman, and Ohanian. "We all started getting touchy from the stress and lack of productive work," Swartz wrote. "We began screaming at each other and then not talking to each other and then launching renewed efforts to work together only to have the screaming begin again."[41] Swartz retreated further into himself, avoiding work and his colleagues, while Huffman effectively assumed the responsibilities of Reddit's only full-time programmer. "The situation was so toxic we were, like, 'This is not gonna succeed; we should just sell while we can,'" Huffman remembered.[42] At times, it felt as if Reddit might implode before the Condé Nast deal was consummated.

Finally, the sale went through on October 31, 2006. "Good news for the Reddit folks. And maybe a very very smart move by CondeNet," enthused tech commentator David Weinberger.[43] "I happen to think that the kind of thing Digg and Reddit and Netscape.com are doing makes a lot of sense as part of a media property of some kind," commented the blogger Mathew Ingram. "No word so far on whether the rumoured price of $65-million has any relationship to reality (I would be suprised [sic] if it was anything close to that)."[44] Though the terms of the deal were never publicly disclosed, the most common estimates for Not a Bug's purchase price fall somewhere between $10 and $20 million. Swartz, an equal equity partner in Not a Bug with Huffman and Ohanian, gave Simon Carstensen some of his own share, in thanks for Carstensen's early work on Infogami. Even after that, Swartz almost certainly received a seven-figure payout.[45]

The night their company was acquired, the newly wealthy entrepreneurs celebrated in Harvard Square. Ohanian and Huffman were exuberant, distributing free Reddit T-shirts to pedestrians, flirting with girls wearing Halloween costumes. But Swartz was oddly sullen. He had dressed as a dot-com millionaire, his costume consisting of a steady gaze and an angry expression. As the evening progressed, the Reddit crew migrated to a bar in Harvard Square. Swartz grew more upset. "I didn't want to be here, I didn't want to know these people," he confessed. "I went home instead and watched a show about a serial killer and found myself identifying with the lead."[46]

The Condé Nast deal required the Not a Bug cofounders to move to San Francisco, where they would work from the offices of *Wired News*. Huffman and Ohanian figured—hoped—that Swartz would take his money and depart. Instead, Swartz decided to stick with the company and relocate to California. He was unhappy from his first day of work at the new office: a large, airy space filled with lots of curious, talkative coworkers. It was, to say the least, a material improvement from the apartment where Reddit had been built. To Swartz, it was a velvet coffin. "The first day I showed up here, I simply couldn't take it," he wrote on his personal blog at the time. "By lunch time I had literally locked myself in a bathroom stall and started crying."[47]

Swartz's behavior was more than a little melodramatic. By normal adult standards, nothing was particularly wrong with Reddit. But it was a *job*, and that was what Swartz found so hard to take. Having a job meant subsuming your personal priorities to your employer's, deferring your own ambitions to focus on the perfection of your product. Unfortunately, Swartz didn't believe in the product, and he wasn't good at pretending otherwise. He quickly shifted from crying at the office to avoiding it entirely. He worked from home, when he worked at all. In December 2006, he traveled to Europe to attend a computer conference and didn't inform his colleagues of his plan; his whereabouts were revealed when his photo, as a conference attendee, appeared on the front page of the *Wired* website. After the conference, instead of returning to San Francisco, he went back to Boston, where he retreated to his old apartment and began working his way through David Foster Wallace's novel *Infinite Jest*.

His unreliability could no longer be tolerated. "He came to the office for the third time ever," Steve Huffman remembered. "I was the only employee there. And I turned to him and said, 'Dude, why are you even here?' Just leave, right? That's what we thought would happen."[48] It's doubtful whether Swartz himself even understood why he was there. He'd spent the last two years working against his nature, following other people's directions down unfamiliar paths. And he appeared to have lost his way.

On January 18, 2007, while still in Massachusetts, Swartz posted to his blog a short story that resembled nothing so much as a suicide note. The protagonist was a young man named Aaron with "body image issues" who couldn't stand the sight of himself. The "Aaron" character slowly starved himself to a cadaverous state. His friends deserted him. He grew listless. Eventually, he sought death:

> The day Aaron killed himself, he was awoken by pains, worse than ever. He rolled back-and-forth in bed as the sun came up, the light streaming through the windows eliminating the chance for any further sleep. At 9, he was startled by a phone call. The pains subsided, as if quieting down to better hear what the phone might say.
>
> It was his boss. He had not been to work all week. He had been fired. Aaron tried to explain himself, but couldn't find the words. He hung up the phone instead.
>
> The day Aaron killed himself, he wandered his apartment in a daze.[49]

Later, Swartz insisted that the story was entirely fictional and chided those who saw it as a cry for help. "I was deathly ill when I came back from Europe; I spent a week basically lying in bed clutching my stomach," Swartz wrote.[50] "I wrote a morose blog post in an attempt to cheer myself up about a guy who died. (Writing cheers me up and the only thing I could write in that frame of mind was going to be morose.)"

Still, the story alarmed his coworkers enough that they dispatched the Cambridge police to Swartz's apartment to prevent him from killing himself. Swartz's mother called to break the news. "Alexis called the cops," she said. Faced with an impending unpleasant situation he could neither avoid nor control, Swartz responded with what was becoming habitual behavior: he ran. "I just make it," he wrote later. "I see the cops coming in front of my door as I make it to the next block." The cops entered Swartz's building. They knocked on his door. Then, getting no response, they broke the door down.[51]

Swartz was fine, but his relationship with Condé Nast and his Reddit colleagues was beyond repair. "I am presented with a letter accepting my resignation," Swartz wrote in January 2007, a few days after posting his suicide story. "I am told to collect my 'personal effects.' A woman from HR politely escorts me from the premises. She never says that she is escorting me, but she does stand behind me wherever I go. I think I am supposed to leave.

"I leave. The sun is shining brightly. It's a beautiful day."[52]

SWARTZ'S departure from Reddit roughly coincided with the beginning of his first serious romantic relationship. He moved in with the writer and artist Quinn Norton, whom he'd met at the Emerging Technologies conference in 2002, when he was fifteen and Norton was in a long-term polyamorous relationship with two men. Now Swartz was an adult and Norton was single. She was also thirteen years older than him, but Swartz had never had a problem with wide age disparities. They began as roommates. They soon became more.

Unchained from corporate life, in love for the first time ever—"[falling in love] takes a shockingly huge amount of time!" he noted on his blog[53]—Swartz spent the first half of 2007 reflecting on his life and his future, and continuing to clarify his objections to Silicon Valley culture. The "animals" of his generation were being incentivized to use their talents on websites that were "the mental equivalent of snack food," he wrote in March 2007.[54] Though computer technology could theoretically be employed to solve many of the world's problems, such an outcome "requires people to sit down and build tools that solve them. Which, as long as programmers are all competing to create the world's most popular timewaster, it doesn't seem like anyone is going to do." On his website, he posted a quote from a writer named Paul Ford on the differences between for-profit and nonprofit endeavor: "If you work for a startup you can fool yourself into believing that the reward will be eternal wealth, but I work for a nonprofit, and the reward is: I did a thing, and I doubt I'll ever do anything like it again."[55] Swartz started to pursue projects that could deliver these sorts of intangible rewards.

In July 2007, Swartz took to his blog to announce a project of this sort. "Early this year, when I left my job at Wired Digital, I thought I could look forward to months of lounging around San Francisco, reading books on the beach and drinking fine champagne and eating *foie gras*," he wrote facetiously. "Then I got a phone call."[56] The phone call was from Brewster Kahle, who wanted Swartz to help him build the ultimate online library.

Since driving his bookmobile to Washington for the *Eldred v. Ashcroft* arguments, Kahle had continued his flamboyant efforts to bring public-domain literature to the public. In 2005, he and others founded an organization called the Open Content Alliance, a consortium of libraries, publishers, and technology companies that agreed to collaborate in digitally scanning books and making them accessible online. "The ancient library of Alexandria collected 75 per cent of all the books of all the peoples of the world in 300 BC," Kahle told the *New Statesman* in October 2005. "Our opportunity is to do that again, but then to one-up the ancients by making it available universally."[57] The Open Content Alliance eventually helped digitize more than one million books—which, while a substantial number, fell far short of Kahle's ultimate ambition.[58] "Brewster is a collector," one of Kahle's collaborators told Elisabeth A. Jones during an interview for her dissertation, "Constructing the Universal Library." "Have you ever met any collectors? Whether they collect maps or cars, they just . . . are acquisitive of STUFF. And that's what he is. And collectors don't stop until they've got all the stuff."[59]

The Open Content Alliance was formed in response to Google Print for Libraries, Google's hugely ambitious attempt to visit academic libraries with advanced scanning machines and systematically digitize every single book in those libraries' collections. It appeared to be the most significant step yet toward making the infinite library a reality, and, indeed, Google today stores many complete public-domain texts on its servers, which can be read and downloaded by anyone.[60] The problem, as Kahle saw it circa 2005, was that Google seemed less interested in building a vast online library than in building a vast online bookstore. "We have been very clear that we want to build a book-finding tool, not a book-reading tool," a Google Print official told the *San Francisco Chronicle*.[61] The project would not create an organized, browsable online library. Instead, it would present limited excerpts that would direct Web users toward retailers where they could purchase the books in question.

The Open Content Alliance wanted to offer an open alternative to Google Print. "As the project proceeded, Kahle's drumbeat of Google criticism grew steadily louder and more strident," wrote Jones. "By 2007, his rhetoric about the company had begun to consistently position it—literally—as an Orwellian information overlord, hungry to absorb and privatize not only the entire library world, but the entire information ecosystem."[62] In a bid to thwart Google's tentacular machinations, Kahle came up with Open Library, another hugely ambitious project that proposed the creation of a separate webpage for every single book that had ever been published. Kahle asked Swartz to help him build it. The site would contain identifying information on each book, and, if available, a link to an online version of the book that a curious reader could browse for free.

Swartz came aboard to build the site's user interface and supervise its metadata implementation. "Our goal is to build the world's greatest library, then put it up on the Internet free for all to use and edit," Swartz wrote on his blog when the site launched in July 2007. "Books are the place you go when you have something you want to share with the world—our planet's cultural legacy. And never has there been a bigger attempt to bring them all together."[63] For free, anyway.

IN October 2007, a year to the day that Not a Bug had been acquired by Condé Nast, Swartz announced on his blog that he had started writing a novel. Called *Bubble City*, it concerned the start-up scene in San Francisco; he promised to post the chapters to his blog as he finished them. "A novel is disguised memoir, as anyone who is friends with a novelist can readily discern," wrote Swartz. "And, like a memoir, writing them can be therapeutic, exorcising a writer's demons, while hopefully being more entertaining."[64]

The novel was set at a small news-aggregation start-up called Newsflip, a barely disguised version of Reddit. The protagonist was a Swartz stand-in named Jason: a new Newsflip hire with a true heart, a keen mind, and terrible social skills. "Curiosity and courtesy were behaviors he had worked hard to learn to imitate so that others didn't find him too strange," Swartz wrote, "but he did his best to make sure other people took no more than a couple hours of his time."[65]

The plot, such as it was, involved Jason's discovery that the code that ran Newsflip had deliberately been corrupted. The site was supposed to work like this: A user logged in, and Newsflip tracked his activity. Eventually, the site took the information it had compiled about the user's habits and preferences to surface the sorts of stories that the user would want to read. But some malign entity had altered the code to show users the stories of the entity's choosing.[66]

That malign entity turned out to be Google, which specialized in projects that appeared to be for the public's benefit, in keeping with its unofficial corporate motto: "Don't be evil." In *Bubble City*, Swartz opined on this hollow-hearted promise. "Don't Be Evil was some hacker's PR ploy that got out of hand," Swartz wrote. "Paul Buchheit, the guy who made Gmail, suggested it in an early meeting and Amit Patel, another early Googler starting [*sic*] writing it on whiteboards everywhere. A journalist saw it and the rest was history—but don't be mistaken, it was never official corporate policy."[67]

How could it be? *Moral Mazes* had taught Swartz that companies cared first and foremost about their own survival and self-perpetuation, and evaluated their business strategies based on those criteria. If it was more profitable to be good than evil, then a company would be good. If it was more profitable to be evil than good, then a company would be evil.

As a teenager in Chicago, Swartz maintained a blog about Google's products and internal affairs, which proved so popular that he had been invited to Google's Silicon Valley headquarters for a tour and a visit. By the time he turned twenty-one, his enthusiasm had curdled. With a heavy hand, Swartz depicted Google in *Bubble City* as a ruthless, all-powerful organization that would stop at nothing— not even murder—to get its way, and its employees and partners as weak-willed collaborators who were forever devising justifications for the compromises forced upon them. As one character in the book put it, "There were a lot of other evils out there to fight, and if working with Google made him more effective at fighting those, wasn't it, on balance, a good thing? And the free food was nice too."[68]

Apathy and amoral corporate solipsism ultimately lay at the tale's heart of darkness. Compromise and rationalization were attitudes that stained the soul. *Bubble City*, which Swartz never finished, breaks off with Jason on the run after discovering Google's secret, while Google invests all its power and energy to track him down. "'So what do you do when you find him?' Wayne asked. 'The usual,' Samuel said as if it was obvious. 'Co-opt or destroy.'"[69]

Swartz published eleven chapters of *Bubble City* before becoming sick with ulcerative colitis. His odd eating habits—he favored bland, white foods, such as plain rice—were efforts to manage a lifelong medical condition that was debilitating when it flared up. "Huge pains grind through my stomach, like it's trying to leap out of my body," Swartz wrote on his blog at the time. "Food is always followed by pain, followed by running to the bathroom."[70] He juxtaposed this description of his digestive ailment with an account of what it felt like to be depressed. "Surely there have been times when you've been sad," he wrote. "Perhaps a loved one has abandoned you or a plan has gone horribly awry. Your face falls. Perhaps you cry. You feel worthless. You wonder whether it's worth going on. Everything you think about seems bleak—the things you've done, the things you hope to do, the people around you. You want to lie in bed and keep the lights off. Depressed mood is like that, only it doesn't come for any reason and it doesn't go for any either." It's hard to say which condition sounds worse.

Swartz no longer had time for ambivalence, or to make foreign suns the center of his universe. On his blog, he described how alienated he felt in San Francisco: "When I go to coffee shops or restaurants I can't avoid people talking about load balancers or databases. The conversations are boring and obsessed with technical trivia, or worse, business antics."[71] Furthermore, he hated the noise of the city, which was inconducive to serious, sustained thought and concentration. The place was filled with people saying things they didn't believe, that they *couldn't* believe, as if, by some extraordinary mass delusion, they had collectively confused the inessential with the elemental, noise with signal.

In March 2008, Swartz traveled to Calgary to address a group of young Canadian leaders at the Banff Forum, an annual festival of conventional wisdom for the middlebrow elite attended by people with expensive business cards that suggested nothing whatsoever about emergent cultural anarchy. The topic of Swartz's panel was "The Internet and Mass Collaboration," and the discussion focused on their implications for politics, business, and rentier capitalism.

Swartz stated what he believed. Free culture wasn't a business model in disguise. It was a revolution:

> The rhetoric often suggests that some magical force of "peer production" or "mass collaboration" has written an encyclopedia or created a video library. Such forces do not exist; instead there are only individual people, the same kind of people who drive everything else.
>
> The power is that these people are collaborating. But they are collaborating because they have come together to form a community. And a community works because it has shared values. But here's the thing: these shared values are profoundly anti-business. [Laughs from the audience.] I mean, look at Wikipedia. This is a group who wakes up every day and tries to put the encyclopedia publishers out of business by providing a collection of world knowledge they can give away to everyone for free.
>
> If you want someone to do your company's work for you, finding a well-organized online community with strong anti-business values seems like a bad idea. [Laughs.]
>
> So what do you do? I have a friend who is even more brash than I am and when anyone asks her for business advice she tells them simply: Well, in the future, your servants are going to rise up and eat you. So, invest in toothpicks.[72]

Upon leaving Stanford, Swartz had called himself a capitalist. By early 2008, he had tossed that self-portrait into the dustbin of history. That spring, Swartz prepared to return to Cambridge, Massachusetts, where he would busy himself biting the hands of the business world that had once tried to feed him. His days in Silicon Valley were behind him forever.

7

GUERILLA OPEN ACCESS

As long as man has lived in society, he has yearned to escape that society and run off to live in a cave. This dream is usually abandoned once the would-be escapee realizes that caves are uncomfortable and that society is where his things are. The material comforts of civilization are magnetic, indeed, and those few individuals who do escape society's pull are almost always drawn by some more powerful force, love or religion or misanthropy or madness, occasionally all at once.

Faith first brought Paolo Giustiniani to the Caves of Massaccio in 1520, faith and disaffection.[1] Born to a noble Venetian family, Giustiniani traded his finery for a friar's habit and joined the Camaldolese, a religious order founded in 1012 by the monk Romuald, who counseled his followers to "put the whole world behind you and forget it."[2] But such exhortations are easier to address than to follow—the world has a way of intruding, even in a cloister. Giustiniani found monastic life insufficiently contemplative, and his fellow monks insufficiently detached from the outside world. When his brethren balked at his suggestion that they resuscitate Romuald's rule, Giustiniani left the community, in search of a spot where he and others could live apart in silence and self-abnegation, contemplating the wickedness of the world and the glories of God, in hopes that the latter would soon obviate the former.[3]

He found his hermitage in a canyon near Cupramontana, in central Italy, where earlier travelers had carved a warren of caves into the rough tuffaceous rock. The caves offered shelter and silence; the canyon was sylvan, serene. In a letter to a friend, Giustiniani noted that he was forced to write "on my knees, not even having a stool."[4] But if you wanted to save the world, you had to first be willing to forsake it. Giustiniani moved in. Others soon followed. For centuries, the monks lived there in the transcendent quiet.

Then modernity came, and with it came disrepair. As went the monastery, so went the friars; vocations, like buildings, can grow worn and weathered.[5] The few remaining monks dispersed between the First and Second World Wars, and today the entire property is deconsecrated and privately held, used as a retreat and conference space by secular groups that also seek to escape society, if only for days at a time.[6]

Though it has evolved from its original purpose, the hermitage still stands as a monument to radical idealism, a physical reminder that we have a choice, that we are not bound inextricably to social convention. It takes courage to opt out of the world, its comforts and vices, in pursuit of less tangible rewards, but it can be done. Uncomfortable as caves may be, we can eventually learn to live in them. If society makes no sense, we can always stop listening. If the world disappoints us, we can leave it behind.

WHEN he came to Cupramontana in the summer of 2008, Aaron Swartz had already made his choice. Like Giustiniani, he had known great wealth and had deemed it curiously insubstantial. "A friend told me to be sure not to let the money change me," Swartz wrote on his blog in November 2006, less than a week after Condé Nast had purchased Not a Bug. "'How could it possibly do that?' I asked.

'Well, first you'd buy a fancy new car.' 'I don't know how to drive.' 'Then you'd buy a big house in the suburbs.' 'I like living in small apartments.' 'And you'd start wearing expensive clothes.' 'I've worn a t-shirt and jeans practically every day of my life.' "[7]

The brass ring he had grabbed turned out to be an anchor, leaving him mired in other people's priorities. "Let's say you want to make a difference in the world. You can learn a skill and go into a profession, where you get bossed around and told exactly what to do by people more powerful than you," Swartz wrote two months after leaving Reddit. "Want to actually make a difference? You'll have to buck the system instead of joining it."[8]

"The system," in all its incarnations, that vague authoritarian stronghold of imprecise menace and organizational inefficiency, had long been Swartz's primary antagonist. Organized schooling, traditional employment—all of these existed in opposition to Swartz's brand of restless, polymathic nonconformism. Systems exist to perpetuate themselves, so they naturally discourage deviance and promote pliant, normative behavior. "The system," in Swartz's understanding, was designed to sustain those who controlled it and to marginalize those who were unwilling or unable to comply.

Before college, Swartz read a book called *Understanding Power*, a series of transcribed discussions with the MIT linguist Noam Chomsky in which Chomsky analyzed and explained the ways in which political power is wielded, acquired, and guarded. Chomsky's theme was the system—and, particularly, defenders in the mass media. A political logic determines everything that does and does not appear in the press. By design the mass media validates and legitimizes the system while excluding unfamiliar perspectives.

Understanding Power resonated with Swartz like few other books he'd read. "The Book That Changed My Life," Swartz called it in a blog post from May 2006. "Reading the book, I felt as if my mind was rocked by explosions. At times the ideas were too much that I literally had to lie down. . . . Ever since then, I've realized that I need to spend my life working to fix the shocking brokenness I'd discovered."[9] At age twenty-one, he was starting to do so in earnest.

Years after he first read Chomsky's book, Swartz thought he understood power pretty well. Knowledge is power. Therefore, free, unimpeded access to information is an inherently political issue, not just a slogan to be monetized or a memorable line in a presentation to a venture capitalist. Open data, attractively presented, can promote organizational transparency. Open access to culture and scholarship can expand perspectives and open minds. By sharing knowledge with people, you can empower them to better apprehend the world and to act on that new apprehension.

In June 2007, Lawrence Lessig also decided to buck the system. On his personal blog, Lessig announced that he was going to stop working on copyright issues and instead shift his focus to political corruption. Lamenting that "our government can't understand basic facts when strong interests have an interest in its misunderstanding," Lessig characterized "corruption" as the product of excessive money in politics.[10] Take the Sonny Bono Act debates, for instance. Congress passed a copyright term extension that enriched the entertainment industry at the expense of the public domain—a decision, Lessig implied, that stemmed from legislators prioritizing the interests of campaign funders over those of the general public. Policymaking in the United States was too often an exercise in willful ignorance. "I am someone who believes that a free society—free of the 'corruption' that defines our current society—is necessary for free culture, and much more," Lessig concluded. "For that reason, I turn my energy elsewhere for now."[11]

During a conference on open government in December 2007, Lessig inquired whether there was any sort of community resource for "scrapers"—people who used the Internet to download large data sets.[12] Swartz decided to create one and launched a website called theinfo.org. "This is a site for large data sets and the people who love them," Swartz wrote by way of introduction.[13] Swartz loved them more than most. Concurrent with his work for Open Library, Swartz had crawled the Google Books archive and downloaded approximately 530,000 public-domain texts.[14] In collaboration with the Stanford law student Shireen Barday, he downloaded the entire Westlaw database of law-review articles and analyzed the corpus to assess the outcomes of corporate-funded legal research. He had launched a website called watchdog.net, which sought to "make government data accessible and useful"[15] by compiling freely available and politically relevant data sources—voting records, campaign finance disclosures, demographic information—and presenting them in a way that encouraged visitors to take political action.[16] If "the system" relies on institutional opacity to conceal its aims and to consolidate its grasp on power, one way to buck the system, Swartz seems to have believed, is to reveal the information that it actively keeps hidden.

"Getting 'real information' to people on the World Wide Web is 13-year-old Aaron Swartz's job," the *Chicago Tribune* had written in 2000, after Swartz launched his proto-Wikipedia.[17] The job remained the same, except now Swartz was working on a bigger scale. Initiate a discussion about "real information" and the Internet, be it on a blog, mailing list, message board, chat client, social networking site, or tin-can telephone, and you could count on Aaron Swartz eventually joining the conversation. He traveled the world evangelizing on these topics, like a man touched with ideological tinnitus, unable to escape the sound of social dysfunction and desperate to make others hear the ringing in his ears.

In July 2008, Swartz traveled to Cupramontana as the guest of the international library consortium Electronic Information for Libraries (EIFL), which works to provide libraries in developing countries with access to the same resources as their first-world counterparts. EIFL booked the hermitage for a two-day board meeting and "visioning retreat," and had invited Swartz to come help the group envision its ideal future.

Norton accompanied Swartz, and they both stood out in the middle-aged crowd. Swartz was by far the youngest attendee, but he found it easy to fit in among the graying librarians who shared his concerns. Over the course of the weekend, the EIFL discussants sat and talked about the issues dear to them all: the fate of open information on an increasingly profit-driven network and the role of the Internet in a world governed by fear and willful misunderstanding.

That's what Cupramontana was for, after all: the contemplation of serious, world-historical questions. But when Giustiniani and the other White Friars arrived at the caves, they had reached the end of their journeys. Although he didn't know it yet, Aaron Swartz's journey was just beginning.

OPEN *access* is an anodyne term for a profoundly transformative idea. Advocates argue that academic research should be made freely available to the world at the time of publication, and that access should not be contingent on an individual's or institution's ability to afford a subscription to a given journal or database. Academic authors do not usually write for profit; rather, their work aims to augment the common store of knowledge. What's more, since the government often funds their research, it's not a stretch to claim that the fruits of that research should belong to the public. So why should this material be subject to the same access restrictions as a mystery bestseller or a Hollywood film? As with many other inexplicable policies, the blame belongs to a vestigial middleman.

When a university professor finishes a research project, she typically records her results in an academic paper, which she submits for publication in a peer-reviewed journal. ("Peer review" is an academic quality-assurance measure in which the author's work is scrutinized by other scholars in the author's field.) These journals—the reputable ones, at least—operate via volunteers, with authors, editors, and peer reviewers all working for free. Nobody gets paid, or expects to get paid, except the publisher. In exchange for the publisher's services, which include coordinating the publication and peer-review processes, formatting, and distribution, the author concedes the copyright to her article in perpetuity. It's a simple trade: the academic publisher assumes the financial risk of preparing and distributing an esoteric work for which there's a limited audience and in exchange retains all the profits that might come from its sale.

In commercial trade publishing, publishers realize profit by selling a book for a relatively low price to a wide audience. Since no wide audience exists for academic papers, academic publishers realize profit by selling them at high prices to the few entities who can't do without them—libraries and scholars, mostly—which renders these papers functionally inaccessible to the casual or impoverished user.

Today, several multinational publishing conglomerates, such as the Anglo-Dutch company Reed Elsevier, dominate the academic-journal business. Tens of thousands of scholarly journals exist, and since the 1970s their subscription prices have risen at a rate higher than the rate of inflation, leading to what librarians have dubbed the "serials pricing crisis." These journals are expensive, especially in the so-called STEM (science, technology, engineering, math) fields—yearly subscriptions to specialized STEM journals can cost more than $10,000[18]—but academic libraries are, more or less, compelled to subscribe. Every professor thinks that *his* field of study is the most important and expects to find his specialization's academic journal on the library's shelves. Thus, many academic libraries wind up spending the bulk of their yearly acquisitions budgets on journal subscriptions.

That's assuming that a library has a meaningful acquisitions budget at all. Many of them do not—or, at least, don't have much of one. This plight is especially common in poor, underdeveloped countries, where librarians have enough trouble keeping their computers on, let alone keeping up with the latest research in a thousand microdisciplines. The result is an ever-widening gap between rich institutions and poor ones: wealthy Western universities can keep their libraries current with all the latest research; less fortunate schools simply go without.

The open access movement emerged in the early 1990s, when librarians and researchers realized that the Internet had the potential to transform academic publishing and possibly solve the serials pricing crisis by shifting the balance of power away from publishers and back to scholars. Online content distribution could reduce the physical production costs that publishers cited to justify their journals' high prices. Publishers could let underresourced clients access proprietary material online and incur no direct financial loss by doing so. Academics might even choose to eliminate the middleman and simply publish their research online, for free. "In the old, Faustian days, the reluctant choice was to accept the Faustian pact (of allowing access to a work only to paid ticket-holders) because that was the only way to reach an audience AT ALL," the open access pioneer Stevan Harnad wrote in 1994. "But now that there is another option, it's time to rethink all of this."[19]

The first significant academic open access project, arXiv, began in 1991, when the physicist Paul Ginsparg created an online repository for preprints of research papers in his field. Before publishing their articles in an actual scholarly journal, academic authors will circulate drafts among their colleagues worldwide, soliciting comments, critiques, and suggestions. ArXiv streamlined that often inefficient process. By uploading their preprints to the archive, authors could request and receive feedback much, much faster than ever before. There was a side benefit for readers who couldn't access those finalized articles in a scholarly journal: the preprint drafts remained on arXiv in perpetuity, free for all to access.

Ginsparg's growing preprint archive augured great changes in the academic-publishing world. "Starting 10 years ago, we no longer needed publishers to turn our drafts into something that had a polished superficial appearance," Ginsparg wrote in 1994. "Starting more recently, we no longer need them for their distribution network—we have something much better."[20]

In 2002, the movement formally promulgated its principles in a document called the Budapest Declaration, which, like the unwritten Robert Ludlum novel of the same name, is a thrilling tale of international intrigue and one-world politics. The authors argued that easing access to scholarly journals would "accelerate research, enrich education, share the learning of the rich with the poor and the poor with the rich, make this literature as useful as it can be, and lay the foundation for uniting humanity in a common intellectual conversation and quest for knowledge."[21]

Two similar documents followed the Budapest Declaration: the Bethesda Statement, in June 2003, and the Berlin Declaration, in October 2003. All three emphasized the immense public benefit of easing access to scholarly research and exhorted publishers to embrace open access policies. But the large academic publishers had no immediate incentive to change their business models; in fact, doing so would diminish the value of their copyrights and reduce their profit margins. Lacking consistent pressure from stakeholders around the world, they would never change their business practices. For them, the system worked.

During the two-day retreat in Cupramontana, Swartz listened and grew indignant as scholars and librarians from third-world countries recounted their struggles with access restrictions. Though the specifics of each story differed, the outlines were similar: outdated libraries, frustrated researchers, underserved students, and unresponsive publishers. Governments passed laws to support old structures rather than encourage the development of new ones. Publishers relegated information behind paywalls, accessible only to those who could afford the toll. Database providers let long-forgotten books and papers molder in their archives, instead of releasing them to the public domain.

To Swartz, these policies were inefficient, and, worse, irrational—an affront to his utilitarian moral code. While short-term financial logic suggested that the publishers should cling tightly to their copyrights, the world would surely be better served if they relaxed their grip. The publishers' business model seemed fundamentally immoral in the Internet age and, left unchallenged, would ultimately prove fatal to the open, collaborative Web. If the public wasn't rising in opposition to restricted-access policies, it could only be because they didn't understand the stakes. So, one night in Cupramontana, Aaron Swartz tried to elucidate the gravity of the situation as clearly as possible.

He began with a simple, declarative sentence that doubled as a statement of purpose. *Information is power*, he wrote, which is why it should come as no surprise that the status quo would prefer to keep information scarce. *Want to read the papers featuring the most famous results of the sciences? You'll need to send enormous amounts to publishers like Reed Elsevier.*[22]

Had they been invited to Cupramontana, Reed Elsevier and its fellow publishers would have disputed Swartz's accusation that they were hoarding the world's cultural heritage to benefit the rich at the expense of the poor—or, indeed, that anything was venal about their buy-low, sell-high business model. Reed Elsevier profited by publishing academic journals and selling them for a fee. Its digital activities were neither particularly nefarious nor unprecedented. The company was simply using the Internet to develop another distribution channel for its products.

In 2003, the Downhill Battle activists had argued that the rise of the iTunes digital music store discouraged musicians and record companies from experimenting with new distribution models, ones that might better serve both musicians and consumers. An analogous interpretation can reasonably be applied to academic publishing. For Reed Elsevier and its competitors to try to impose an old model on an emerging medium, as if the Internet were just a printing press with a new paint job, wasn't just a matter of ignorance or inertia, but an act of willful obstruction.

There are those struggling to change this, Swartz continued. He was one of those people. So were the others at the EIFL retreat, and all the far-flung open access partisans who had *fought valiantly to ensure that scientists do not sign their copyrights away but instead ensure their work is published on the Internet, under terms that allow anyone to access it.* But change wasn't coming fast enough. With every day that passed under the current state of publishing and copyright, more and more material was being unjustly withheld, accessible only to those who could afford it, denied to those who could not.

It's called stealing or piracy, Swartz wrote, *as if sharing a wealth of knowledge were the moral equivalent of plundering a ship and murdering its crew. But sharing isn't immoral—it's a moral impera-tive. Only those blinded by greed would refuse to let a friend make a copy.* Any reluctance to embrace free culture is a function of greed: for Swartz, the moral equation was often that simple. During his work with Open Library, for instance, Swartz hoped to utilize existing on-line catalog records that had been created by a pioneering library consortium called the Online Computer Library Center (OCLC). OCLC was the first group to digitize massive amounts of card-catalog information, and its database, WorldCat, today contains almost 340 million bibliographic records itemizing the collections of libraries around the world. OCLC is inarguably the world's largest and most important library catalog, and Swartz came to believe that it did not want Open Library to challenge its supremacy. In November 2008, OCLC changed its terms of service to now stipulate that users were forbidden to use WorldCat data for any project that "substantially replicates the function, purpose, and/or size" of OCLC's database.[23]

The consortium's apparent power play infuriated Swartz. "As servers have gotten cheaper, it's become easy to do for free the things OCLC charges such outrageous amounts for. But OCLC can't have that," he argued. "So they're trying to stamp out the competition."[24] He couldn't abide the logic of profit, whether it was advanced by OCLC, Reed Elsevier, or any other entity. After railing against *large corporations* that were unwilling to work toward open access, Swartz came around to his point, his call to action:

> There is no justice in following unjust laws. It's time to come into the light and, in the grand tradition of civil disobedience, declare our opposition to this private theft of public culture.
>
> We need to take information, wherever it is stored, make our copies and share them with the world. We need to take stuff that's out of copyright and add it to the archive. We need to buy secret databases and put them on the Web. We need to download scientific journals and upload them to file sharing networks. We need to fight for Guerilla Open Access.
>
> With enough of us, around the world, we'll not just send a strong message opposing the privatization of knowledge—we'll make it a thing of the past. Will you join us?

He called the document the "Guerilla Open Access Manifesto," which, in retrospect, was perhaps nomenclaturally unwise. *Manifesto* connotes instability and political upheaval, the rise of people with nothing to lose but their chains; *guerilla*, for its part, brings to mind barbate insurgents in berets, toting Kalashnikovs through some fetid jungle. The title certainly suggested that Swartz stood for anything but peaceful, law-abiding resistance. That implication caused discomfort among some open access supporters.

Quinn Norton has claimed that others at the conference also contributed to the manifesto. But Swartz was the only one willing to take responsibility for it, which perhaps answered the question posed in the manifesto's final line. This reluctance to embrace Swartz's cri de coeur persisted as the document spread throughout the open access community. "I thought he was ethically right, but I was unwilling to put my own livelihood on the line with such strong statements," the librarian and former EIFL affiliate Bess Sadler wrote years later of Swartz's polemic. "A librarian who issued a manifesto like that would be unemployable, and that's something that should give us all pause."[25]

Librarians and academics, after all, had to work for and exist within the system they were criticizing. They tended to support incremental measures achieved through indirect means: petitions, consensus building, conferences, visioning retreats. But Swartz, now a freelance idealist, was uninterested in waiting around for systems to gradually reform themselves. Four years after Chomsky's *Understanding Power* first sent him reeling with the vertiginous power of its transgressive political ideas, Swartz was surer than ever that systems existed to be bucked and refuted. If you want to save the world, sometimes you have to be willing to take to the caves.

IN 1991, when both the Internet and Aaron Swartz were still young, a man named Carl Malamud launched a guerilla open access campaign of his own. The International Telecommunication Union, a nonprofit group that promulgates global telecommunications standards, agreed to let Malamud post its nineteen-thousand-page *Blue Book* standards manual on the Internet.[26] The ITU earned about $5 million per year from the manual's sale, which was a good thing for the ITU, if not for the engineers who had to pay the *Book*'s steep purchase price.[27] But now the standards were available online, for free, and they had proved much more popular than anyone had expected.

Malamud, at the time, was a thirty-two-year-old consultant and writer of professional reference books; his work required access to standards documents, and he was routinely frustrated at their inaccessibility and excessive cost.[28] Treating technical standards as if they were trade secrets only served to slow their global adoption. As he put it, "Standards have to be widely available or the standards are irrelevant."[29]

So Malamud approached the ITU and volunteered to take their standards, convert them from the group's archaic internal format into something more modern, and post them online. (Malamud billed himself as a founder of something called the Document Liberation Front, which probably should have given pause to the ITU.)[30] The ITU, as an institution, still labored under the misimpression that the Internet was a niche medium used mostly by academics. The body agreed to comply with what it thought of as Malamud's experiment, with no expectation that the *Blue Book* would find any significant online audience.

They couldn't have been more wrong. "The academic toy, to the bureaucracy's horror, turned out to have over seven million people," Malamud wrote. "They were shocked to see hundreds of thousands of ITU documents being accessed by thousands of people in dozens of countries."[31] The ITU, envisioning that $5 million in annual sales evaporating into networked oblivion, hastily ended the project and attempted to reassert control over the *Blue Book*. Officially, the ITU stated that the "successful experiment has now served its purpose."[32] Writing in the trade publication *Communications Week*, a peevish Malamud suggested some ulterior motives behind the reversal: "If we gave away the standards, there would be fewer jobs at the ITU. There would be less control over distribution and more pressure to start responding to the realities of engineering in the rest of the world."[33]

But Malamud had made his point. Though observers might dismiss the Internet as a novelty, the network had immense potential as a conduit for useful information. Later, Malamud compiled the results of his work with the ITU standards into a curious book called *Exploring the Internet: A Technical Travelogue*, in which he examined the ease with which the standards spread as well as the folly of trying to suppress data online. "The Internet is here," wrote Malamud, "and it is not an academic toy."[34]

Since then, Malamud has dedicated his life to demonstrating all the ways the Internet will change the world, once the world gets out of the Internet's way. In 1995, Malamud was the first person to broadcast live online audio feeds from the floor of Congress. In 1996, in an effort to raise public awareness of the Internet and its potential as a democratic tool, Malamud organized an online world's fair, featuring "pavilions" from countries around the world. "Previous world's fairs have been very effective at exposing millions of people to new technologies," Vice President Al Gore wrote in a letter endorsing the project. "A World's Fair for the Information Age will give people a look at the not-too-distant future—a future in which all of us will be connected by a global network of networks."[35]

Malamud's Internet evangelism was inextricable from his open-data advocacy. "The Internet is a powerful force for democratic values and for an informed, responsive government," he wrote in 1993, and he spent much time making the case to skeptical bureaucrats who did not take naturally to the Internet.[36] The earliest federal websites lacked substantive information and functioned primarily as promotional brochures. More meaningful Internet adoption was inhibited by officials who dismissed the network as little more than a glorified party line for eggheads. In 1993, after listening to a representative from the Securities and Exchange Commission explain that the SEC wasn't making corporate filings freely available on the Internet because Internet users weren't "the right kind of people," Malamud indignantly blurted that he, for one, thought "the American people were the right kind of people."[37]

While Frank Capra would have applauded Malamud's retort, the SEC official was somewhat less enthusiastic. "My input was not appreciated," Malamud noted, but that was nothing new: a prophet, after all, is rarely appreciated in his own time. But the promised land was nigh, so to speak, and Malamud was determined to lead the federal government to it, by means of good example, moral suasion, or, if necessary, blunt force.

He used all three tactics with the SEC. Each quarter, the SEC requires all US public corporations to file reports that offer an accounting of their recent financial activities; these companies are also required to disclose other changes of state, such as insider stock transactions or executive turnover. In the late 1980s, the SEC hired private contractors to manage its electronic filing system, which was called Electronic Data Gathering and Retrieval (EDGAR). One of those firms, Mead Data Central, took the EDGAR filings, digitized them, and charged institutional subscribers approximately $138,000 per year to access the data online. (An overnight data-tape delivery service was also available, for the comparative bargain price of $78,000 per year.)[38]

These documents have immense public value. They are the primary sources that journalists, investors, public watchdogs, and others use to gauge public companies' financial health, evaluate their business strategies, and detect potential corporate malfeasance. The good stuff appears in the SEC filings, not because these companies *want* to reveal their data but because they *must*. When Malamud and others started wondering publicly why the SEC didn't just put the disclosure documents online for free, one of the responses they received was that the SEC was worried that doing so might ruin the market for the services provided by their commercial data contractors.[39]

Existing contracts and relationships with private firms had fostered the sort of disconnect between the government and the public that Lawrence Lessig would later decry. Though the SEC database was comprised of public data, the agency persisted in treating it as proprietary information. "The SEC database is not a product, but the way that investors are informed of the status of public corporations so they may direct their investment dollars to the proper place," Malamud wrote in 1997. "Large government databases are not products; they are the very fuel that makes an information economy function properly."[40]

Since the SEC filings *were* public data, once they were purchased from Mead they could be redistributed freely and legally. Theoretically, nothing could stop Malamud from subscribing to EDGAR and then posting every document for free online. And that's just what he did. He obtained a grant to purchase the data tapes of all SEC filings—the $78,000 option—and, starting in 1994, posted those filings on his website.[41] Gradually, the database found an audience. Then, in August 1995, Malamud announced that he would cease updating the free SEC database in October of that year.

"We weren't running the SEC database because we wanted to be in the database business; we wanted the SEC itself to provide free public access to its information," Malamud later related.[42] The announcement had its intended effect. Eventually, facing public outrage over the vanishing resource, the SEC conceded and agreed to take over Malamud's project.

Malamud spent the next fifteen years deploying similar tactics with other, similarly recalcitrant government agencies and offices. Though his targets changed, his position remained constant: public data belongs to the public, and the government should not unnecessarily restrict the public from accessing that data. In a 2001 *New York Times* article about Malamud, a friend, Paul Vixie, spoke approvingly of the archivist's work: "Often, he succeeds at what I think his primary goal should be, which is to change the world."[43]

In 2007, with funding from Google, Yahoo, the Omidyar Network, and other donors, Malamud launched his biggest shot yet at effecting this change. He established a nonprofit website called public.resource.org, which was intended to host public-domain data sets of all sorts.[44] Not long thereafter, the intrepid archivist announced yet another guerilla data campaign, this time to be waged against a federal database called Public Access to Court Electronic Records, or PACER.

The database, which was created by the US federal court system in 1988, is a comprehensive online archive of federal court documents: millions of motions, transcripts, affidavits, all searchable and downloadable. PACER is an invaluable resource for researchers, who, rather than having to rummage courthouse archives for the files they need, can access that material from the comfort of their own homes. This convenience comes at a cost, though: PACER users must pay ten cents for every page they access. (The fee is only assessed if more than $15 worth of charges is accrued in a given quarter.)

Ten cents per page is not an exorbitant price, in either absolute or relative terms. The statutes supervising Court Fees for Electronic Access to Information, 28 US Code § 1913, specify that the court system can "prescribe reasonable [access] fees," but "only to the extent necessary" to "reimburse expenses incurred in providing these services."[45] In practice, though, the yearly revenue generated by PACER exceeds maintenance costs.[46] The surplus income has been spent on other things: installing flat-screen monitors and speakers in federal courtrooms, for instance.[47]

The fees and the surplus income notwithstanding, the court records *are* online and accessible, and that is a legitimate accomplishment, especially considering other federal agencies' reticence to digitize their data. "PACER is the greatest technological achievement in the court system in the last twenty years," a spokesman for the US Courts would later boast to *Wired*.[48]

Malamud disagreed. "The PACER system is the most broken part of our federal legal mechanism," he told *Wired*. "They have a mainframe mentality."[49] To Malamud, PACER was "a classic old-fashioned 'big iron' project with tons of big computers and loads of contractors," which inherently generated bloat and inefficiency.[50] There were easier ways to get federal court records to the people who needed them, he felt. And there ought to be a way to make all that PACER data available online for free.

In 2007, PACER announced that, for a limited time, it would offer completely free access to its database.[51] (At the time, PACER charged only eight cents a page, not ten.) The free access was available only at sixteen federal depository libraries across the United States—"that's one library every twenty-two thousand square miles, I believe," Malamud quipped[52]—and researchers would have to physically visit these libraries to take advantage of the offer. But once you were logged in to one of those computers, you could download as much data as you wanted.

Malamud posted a message on his website announcing what he dubbed the Pacer Recycling Project and soliciting volunteers for an informal "thumb drive corps." Malamud enjoined members of the corps to visit the depository libraries, download PACER records to portable flash drives, and then "recycle" that material by uploading it to resource.org, where it would live in perpetuity as a free alternative to PACER. "Is this legal?" Malamud asked, before answering his own rhetorical question. "You betcha! These are public documents."[53] Theoretically, if enough people answered Malamud's call, they could siphon the entire PACER database.

Malamud didn't expect that to happen, though.[54] The "thumb drive corps" was more a symbolic initiative than anything else; an example of all that *could* be done if people only exerted the will. On September 4, 2008, he got an e-mail from someone who did. "Do you have any guidelines for the thumb drive core [*sic*]?" inquired Aaron Swartz. "Any particular things we should be sure to capture with the pacer docs?"[55]

AARON Swartz had long admired Carl Malamud. In 2002, when he was fifteen, Swartz spotlighted Malamud on his blog as "today's featured superhero," lauding him as "an unstoppable technical and social hacker."[56] That same year, he asked Malamud to grant him the web.resource.org subdomain, which Swartz planned to use to host "information that's useful to the Web community."[57] "It's a mutual admiration society," responded Malamud. "Of course, we'd be more than happy to delegate web to you!"[58]

During the year prior to their PACER collaboration, Swartz had reestablished contact with Malamud, soliciting his advice and assistance with watchdog.net and a slew of other bulk-downloading projects. Swartz had developed several methods of acquiring large data sets. Sometimes he'd purchase them. Sometimes he'd request them directly from government agencies under the Freedom of Information Act. Sometimes, his tactics were less direct.

The free trial of PACER was made available to on-site patrons of the various chosen libraries—the Alaska State Court Law Library, the Sacramento County Public Law Library, the Portland (Maine) Public Library, and thirteen others[59]—but Swartz figured that, rather than sit at a library terminal all day, it would be simpler to deploy a computer program that would download the PACER data remotely and automatically. This method of acquiring databases was quick and easy—especially for Swartz, who so disliked having to ask other people for help—but it also had the potential to greatly annoy the database providers.

In a blog post published in January 2013, the librarian Eric Hellman recalled how, upon meeting Swartz, he took him to task about "how some of his mass-downloading was getting people really upset and could have negative consequences for the things he was trying to accomplish. If he would just ask, I told him, he could have an account for [a database tool] that DIDN'T crash to smithereens when asked for millions of records. And people were working really hard to make the information he wanted free, it just needed some years to make sure the machinery wouldn't collapse. Aaron sounded embarrassed."[60]

Embarrassed though he may have been, Swartz had no intention of changing his ways. This attitude complicated his collaboration with Carl Malamud. Throughout his career as a data liberation activist, Malamud had always taken care to work strictly within the bounds of the law, both as a means of self-preservation and as a way of underscoring a broader point: public data, by law, belonged to the public, and there was nothing illegal about making it public.

This attitude was sensible. Taking care to comply with every federal database's terms of use often proved time-consuming and inefficient. But by doing so, the downloader retained the moral high ground. Shortcuts cast a shadow on conduct and increased the likelihood of governmental scrutiny and suspicion.

The Princeton professor Stephen Schultze had written a simple computer script that was designed to crawl and download PACER; Swartz had helped code the program, and was now itching to deploy it. But the terms of the PACER access initiative did not explicitly authorize siphoning the database remotely, and this made Malamud nervous. "do you have your library's permission/tacit agreement to drain pacer?" he asked.[61] "no," Swartz replied.[62] "sigh. this is not how we do things. :)," Malamud e-mailed Swartz on September 4, 2008. "we don't cut corners. we belly up to the bar and get permission."[63] If Swartz wanted to collaborate with Malamud, he would have to play by the rules.

Swartz gave his assent and then, without telling Malamud, ran the program remotely anyway. He persuaded a friend in California to visit the library in Sacramento and surreptitiously download an "authentication cookie"—a digital keycard, basically—that Swartz could use from home to fool PACER into thinking he was at the Sacramento library. In Massachusetts, Swartz ran the program, and then sat back and watched the files roll in. "We're going to have fun with this," Malamud told Swartz in late September, after Swartz had estimated that he would be able to capture approximately four terabytes worth of PACER records.[64] "awesome. :-)," Swartz replied.[65]

Around the same time, on September 20, 2008, Swartz revisited the "Guerilla Open Access Manifesto" in a blog post promoting the launch of a website called guerillaopenaccess.com. "I realized that the Open Access movement simply wasn't enough—even if we got all journals going forward to be open, the whole history of scientific knowledge would be locked up," he explained. "Talking with others at the [EIFL] meeting, I realized what must be done. If we couldn't get free access to this knowledge, folks would have to take it."[66]

He reiterated those points that morning at a Free Software Foundation–sponsored event called Software Freedom Day, where he delivered an "eye-opening" keynote address on public records and Guerilla Open Access. "[Swartz has] been using free software to make government records and other public domain material easily available and searchable by the public," the Free Software Foundation's blog reported. "He implored us to all call him, if we want to help."[67] As Software Freedom Day neared its end, Richard Stallman delighted attendees with a surprise appearance. Wearing a plaid shirt and khakis, he spoke briefly on the history of the GNU Project and the future of free software and society. "He exhorted us to think of the GNU project's 25 year history as a foundation for the work to come," the FSF blog noted, "and encouraged people to keep pushing for a completely free system."

A week after Software Freedom Day, the government noticed the unusually high number of downloads purportedly originating from the Sacramento County Public Law Library and severed Swartz's access to the PACER database. Swartz's script had taxed the database, eventually crashing it. But Swartz didn't realize that at the time; all he knew was that his log-in attempts now elicited an "Access Denied" message.[68] When Malamud learned that Swartz had been running his crawler remotely despite instructions to the contrary, he told Swartz, "You definitely went over the line, even after I specifically told you I didn't want that to happen on my resources."[69] Then, worse came to worst: fearing a security breach, PACER suspended the trial-access program entirely.[70]

Swartz had downloaded almost 20 million pages from PACER,

which constituted about 20 percent of the entire database. Malamud feared that Swartz's actions might spark some sort of investigation, maybe even a legal action.[71] Automatically downloading PACER wasn't illegal, as far as they believed, but it was certainly unusual, and as Malamud knew well, federal agencies tended to be suspicious of the unusual. "If they want to go after you, I'll shield you as long as I can, but at the end of the day, we'll simply agree that you did what you did," Malamud wrote to Swartz on September 30.[72] "There was not an explicit rule against what you did. It was pretty stupid, I think, but the motive was good."

A few months passed, and no consequences seemed to be forthcoming. In a *New York Times* article about Swartz and Malamud and the PACER incident, John Schwartz reported that a US Courts spokeswoman was unable to comment on "whether there had been a criminal investigation into the mass download."[73] But the FBI had indeed been busy investigating the affair. In a report dated February 6, 2009, the Washington field office of the FBI noted that, thanks to Swartz's actions, "the PACER system was being inundated with requests. One request was being made every three seconds."[74] Wondering exactly what Swartz and Malamud had been up to, the agency initiated an "information gathering phase."

The file that the FBI opened on Swartz contained a précis of his recent activities. It noted his involvement with watchdog.net, and his ambitions of "pulling all information about politics, votes, lobbying records, and campaign finance reports under one unified interface." Swartz's personal website, the FBI observed, "includes a section titled 'Aaron Swartz: a lifetime of dubious accomplishments.'" The FBI agents reported that Carl Malamud had "published an online manifesto about freeing PACER documents," and that the exploits of Malamud, Swartz, and the thumb drive corps had been covered in the *Times* under the headline "Steal These Federal Records—Okay, Not Literally."[75]

Each of these items was innocuous when taken on its own. Taken together, from the government's perspective, they seemed to indicate some sort of overarching nefarious scheme to do . . . something. In February, the FBI sent a car to surveil Swartz's parents' house in Highland Park. On April 14, 2009, an agent called Highland Park hoping to talk with Swartz in person. Swartz wasn't at home, but the FBI agent spoke with his mother, who was spooked enough to send Carl Malamud a frantic e-mail and Twitter message informing him what had happened. ("tell your mother that twitter is *not* the right way to reach me on this stuff :)," Malamud told Swartz.)[76]

Swartz eventually returned the call on his cell phone. "I'm sure you can guess what this is about. PACER," said Special Agent Kristina Honeycutt, in Swartz's telling. "We're interested in sitting down and talking to you about it, more so to just find out exactly what happened, so we can help the US Courts get their system back up." Honeycutt asked if Swartz would be willing to meet at some point soon for a face-to-face conversation. "If it was something bigger than that," she said pointedly, "we wouldn't have called you to ask."[77]

"you shouldn't worry about me. I'm happy to take the fallout if it comes," Swartz wrote to Malamud, aware that Malamud was at the time vying for a job at the Government Printing Office.[78] "understood, but I'm not going to let them hang you out to dry," Malamud responded.[79] The situation never came to that. Swartz's lawyer eventually called the FBI and said that Swartz would agree to meet only if the agency could guarantee that doing so would not work to his detriment. The FBI couldn't make that promise, so Swartz never met with them. The investigation was eventually closed on April 20, 2009. Later, Swartz requested his FBI file and posted the contents online.[80] Though Swartz feigned bravado in his blog post about his FBI file, calling the document "truly delightful," he had been scared witless at the time it was being compiled.

Swartz had spent two years downloading and uploading various data sets in a flurry of shotgun activism, spreading his shot wide, not caring particularly about *which* target he hit. When he had first proposed joining the thumb drive corps, Malamud had advised him not to let PACER distract him from more potentially significant projects such as watchdog.net. Swartz hadn't listened, and now PACER had gone sour and watchdog.net had gone stale. "There seems to be an impression or at least a worry that you've simply dropped watchdog on the floor and the whole thing could be finished very soon," Malamud wrote to Swartz in January 2009. "What's up?"[81] Swartz acknowledged that there would soon be some "personnel reshuffling" at watchdog.net, and Malamud gently chided him for failing to keep his supporters apprised of the changes. "communication is, by far, the hardest part of leadership," Malamud wrote.[82] "no, you're right. i have not been doing a particularly good job keeping up with people lately," Swartz responded.[83] "not a problem, just self-correct," said Malamud. "screwing up is fine, not realizing it is not. :))"[84]

But far from convincing Swartz to curb his ambitions and proceed with more caution, the PACER experience, if anything, drove him deeper into the caves. Swartz's guerillaopenaccess.com website linked to the website of a group called the Content Liberation Front, self-described "guerillas of the open access movement." The Content Liberation Front's website was a simple list of projects, the first of which was the acquisition of expired journals.

"Many online journal sites, like JSTOR, even charge for articles which have entered the public domain," the site said. "If you have copies of such articles, please upload them to archive.org and let us know." But uploading public-domain articles was only the first step: "If you have a bit more skills or time, we suggest liberating entire journal archives from these sites and uploading them to file sharing networks. If anyone does so, let us know, we'll post about it here."[85]

The site urged visitors to send hard copies of databases to its mailing address:

> The Content Liberation Front
> c/o Aaron Swartz
> 950 Massachusetts Ave., #320
> Cambridge, MA 02139
> USA

That was Swartz's apartment, between Harvard Square and Central Square, just down the road from the Massachusetts Institute of Technology.

8

HACKS AND HACKERS

JSTOR, which stands for "journal storage," is an online database of academic journal articles that was conceived in 1993 and launched in January 1997. With complete archival runs of scholarly journals in many academic disciplines available to institutional subscribers in an instant, JSTOR, in many ways, could be considered the incarnate dream of the infinite library. Yet it also exemplifies the failure of that dream to properly materialize.

In his comprehensive history *JSTOR*, Roger Schonfeld recounted the genesis of the service. Academic librarians, facing the serials pricing crisis, as well as perennial budget and storage constraints, had long discussed the notion of a "central lending library for periodicals" that would obviate the need for individual libraries to archive and subscribe to these journals themselves.[1] In 1993, an enterprising fellow named William G. Bowen, the president of the Andrew W. Mellon Foundation and an economist who had studied nonprofit structures, first floated the idea for a digital version of this central lending library. Bowen may have been a dreamer, but he was an extremely well-connected one. Within a year, plans were under way to turn his chimerical concept into JSTOR, which would serve both as a potential solution to the serials crisis and as a "real demonstration that large-scale digital libraries were feasible."[2]

But making JSTOR a reality proved challenging to Bowen and his colleagues. JSTOR officials initially struggled to convince some academic publishers to participate in the project. Certain publishers worried that signing on to JSTOR in the present might preclude them from monetizing their digital backfiles in the future. Other entities, mostly learned societies that published journals in their respective fields, feared that scholars would cancel their learned-society memberships once they could access these journals online from a central database.[3]

To assuage these concerns and others, JSTOR took care to reassure publishers that their participation would not diminish the value of their content. JSTOR would assume all digitization and archival maintenance costs. The service would not be premised on exclusivity, so publishers would be free to make their backfiles available elsewhere, too. Moreover, as Schonfeld noted, JSTOR guaranteed publishers "that participation in the project would bring them no harm—in terms of lost revenues or other concerns."[4]

The reassurances worked. JSTOR launched in 1997 and has expanded ever since, to the delight of the many students and scholars who have come to rely on its vast digital archives. But soon enough, observed Schonfeld, "JSTOR began to behave like a business, with proprietary rights that required protection."[5] And when those rights were threatened, JSTOR did not hesitate to act in self-defense.

IN the early evening of September 25, 2010, a JSTOR employee noticed something strange. The JSTOR website was sluggish: tasks were accumulating and going uncompleted, Web forms weren't loading. At 6:48 p.m., the staffer reported the problem in an e-mail to colleagues with the subject line "website sad."[6] Nobody likes a sad website, especially not those people tasked with keeping it happy. The JSTOR tech team examined the problem and three minutes later identified its cause: someone was bombarding the JSTOR servers with download requests. Hundreds per minute. And those requests were unraveling the system.

The user in question was clearly using a computer program to initiate download sessions in rapid succession and acquire articles from JSTOR's database in the process called *scraping*.[7] These actions violated JSTOR's terms of service—and, of more immediate concern to the employees on duty that Saturday night, they threatened the stability of JSTOR servers in Ann Arbor, Michigan.[8] Soon, other JSTOR staffers chimed in. "Any chance the offending scraper has an IP from the Portland area?" one asked. "We had a tool from Portland State University apologize and admit he was using 3+ PCs to mass download after they went to his house and punched him in the face (if only)."[9] (Note: These e-mails are taken from a store of internal e-mails voluntarily released by JSTOR, from which all names have been redacted. When I use real names, it means that I have been able to independently verify the writer's identity.)

But the activity was coming from the Massachusetts Institute of Technology—and as the night went on, the scraper started to pick up speed.[10]

The JSTOR system couldn't handle these voluminous download requests. Much as a home computer might freeze when launching a dozen programs simultaneously, a computer server can easily stall if hit with lots of requests in rapid succession. It takes a powerful machine to survive such an onslaught, and the JSTOR servers were, apparently, relatively feeble, or at least unprepared.

After some internal debate about what was happening and how to respond, one of the JSTOR staffers sent out the order to "Jack 'em"[11]—that is, to ban the offending MIT IP address from the system. "You mess with the bull," said another, "you get the horns."[12]

And that, for the moment, was that. ("Time of death: 8:56 pm," one JSTOR employee noted.)[13] But by eight the next morning, the scraper had reactivated at a different IP address and resumed downloading.[14] While JSTOR limited the number of articles that a given user could download per session, it did not limit the number of sessions a user could initiate.[15] The MIT scraper had identified this loophole and, at peak activity, had initiated over two hundred thousand download sessions in a single hour—an average of 55.5 new sessions per second.[16] "This is too much activity for the system," one staffer wrote, and JSTOR responded by again banning the offending IP address.[17] Moves and countermoves: the scraper neatly evaded JSTOR's second ban by adopting yet another IP address.[18] This time, JSTOR retaliated by showing the entire school its horns, temporarily banning a wide range of MIT IP addresses. The downloads ceased, and the JSTOR site slowly recovered. Employees of both MIT and JSTOR proceeded to assess the damage.

In an e-mail to Ellen Finnie Duranceau, JSTOR's contact at MIT Libraries, JSTOR user services manager Brian Larsen refrained from characterizing the incident as a hostile attack, noting, "This activity is normally a compromised username and password or a student/researcher unaware of the impact of their activities or that this method of gathering PDFs is in violation of our Terms and Conditions of Use." The method—"robotic harvesting"—was not only prohibited, it was unnecessary. Larsen noted that JSTOR was accustomed to working with scholars who required bulk access to articles for research purposes "and would be happy to do so in this case as well if that turns out to be the motivation."[19]

Duranceau and MIT's information technology department soon determined that the download requests originated on a computer that had logged in to the school's network with a guest account, which meant that MIT could not precisely identify the guilty party. Nevertheless, MIT recorded the offending computer's MAC address—basically, an identification number that is unique to every computer's network adapter, like a fingerprint—and banned that address from the network. Duranceau told Larsen that the harvesting was unlikely to happen again.[20]

Larsen was glad to hear it. In a follow-up e-mail, he emphasized that the incident wasn't just a typical instance of overzealous archival research, but "an extreme case" that had affected the performance of JSTOR's website, which rarely happened. The IP ban successfully stopped the downloads, and Larsen hoped that JSTOR had sent the downloader a sufficiently strong message: "I have no reason to believe we would need to be more heavy handed than that, but, of course, it is possible that it might be warranted at some point. I highly doubt it."[21]

His faith was misplaced. On October 9, 2010, a JSTOR employee sent an ominous e-mail informing colleagues, "The MIT scraper is back."[22] Just as before, the scraper would start a session and download a document, then start another session and download another document, then repeat the process ad infinitum. With its servers suffering under the strain of the scraping and other users' activities affected by these actions,[23] JSTOR, in an unprecedented move, blocked access to its database for the entire MIT campus to maintain its server stability.[24] Or, as one JSTOR employee put it, "MIT went Rambo on us, and we suspended the whole range."[25]

For JSTOR, this was a drastic measure, since it affected all MIT users, not just the overzealous ones. Blanket bans for entire institutions risk eliciting angry screeds from scholars wondering why, exactly, the database had failed them in their hour of need. Access remained suspended for several days before it was finally restored on October 12.[26] Drastic though it may have been, the solution seemed to work. The downloads ceased.

An analysis of the two incidents produced some alarming numbers. In October, the scraper had downloaded 8,422 articles in 8,515 total sessions. In September, however, the scraper had acquired 453,570 articles from 562 different journals over 1,256,249 sessions. "This is an extraordinary amount and blows away any recorded abuse case that I am aware of," Larsen noted.[27]

What could anyone want with that many articles? The extent and pattern of the robotic harvesting indicated intentionality; the scraping clearly wasn't the work of a student or a professor who had fallen down a research hole. Worries mounted after a deep dive into the downloaded content pointed toward a disturbing conclusion. The first document that had been downloaded in the October scraping session was the article "The Mystery of Misspelling" from a 1957 issue of the *Elementary School Journal*. The final article downloaded came from a 1950 issue of the *Elementary School Journal*. After presumably considering and dismissing the possibility that the system had been breached by a nostalgic fourth-grade English teacher, a JSTOR employee stated what appeared to be obvious: "They're clearly going after substantially the entire corpus."[28]

"The entire corpus" referred to the whole of the JSTOR database: more than 5 million articles from more than a thousand academic journals, all of which had been legally licensed and carefully digitized by the nonprofit organization. In September, MIT told JSTOR that a guest had been responsible for the downloads and that the problem was unlikely to recur. But it *had* recurred, prompting questions that MIT seemed reluctant or unable to answer. Who was draining the database? And why? JSTOR officials worried that voracious overseas hackers had downloaded the files.[29] "By doing a simplified Chinese language Google search on 'EZProxy password,' you will find numerous lists with valid authentication information for hundreds if not thousands of schools," one JSTOR employee wrote, implying that unscrupulous foreigners might be siphoning the archives.[30]

A senior JSTOR official reacted with alarm, asserting that the "activity noted is outright theft and may merit a call with university counsel, and even the local police, to ensure not only that the activity has stopped but that—e.g. the visiting scholar who left—isn't leaving with a hard drive containing our database."[31] Another JSTOR employee concurred: "This is an astronomical number of articles— again, real theft (and one can assume willful malfeasance given the use of a robot, etc.). Does the university contact law enforcement? Would they be willing to do so in this instance?"[32]

IN September 2010, Aaron Swartz purchased a new Acer laptop and visited the Massachusetts Institute of Technology, planning to download as many articles as possible from JSTOR. Logging on to the school's network under the alias Gary Host (G. Host, or "ghost"), Swartz played patty-cake with JSTOR's and MIT's tech teams for months before finding a way to access the database without arousing attention.

His actions shouldn't have surprised anyone. If the city of Cambridge had compiled a yearbook of all its residents, Aaron Swartz would surely have been named Most Likely to Try to Download the Entire JSTOR Corpus. Swartz was an ideologue who had spent the past few years not only bulk-downloading large data sets that were inaccessible to the public, but also writing and speaking on the moral necessity of doing so. The JSTOR hack derived directly from the Guerilla Open Access playbook and the Content Liberation Front's to-do list.

In late September 2010, Swartz traveled to Budapest for the Internet at Liberty conference, where he spoke on "online free expression and enforcing ethics & accountability for corporations & governments."[33] At the conference, Noam Scheiber of the *New Republic* reported, Swartz dined with some activists who, with the backing of George Soros, had tried to get JSTOR to make its archives available to the public. But the price had been prohibitive—securing all the necessary copyrights would have cost Soros hundreds of millions of dollars—and Swartz's dinner companions decried "the outrageous sum of money it would take to free up JSTOR for public consumption."[34] Scheiber makes clear that Swartz's companions did not propose any sort of guerilla downloading campaign or suggest that Swartz take matters into his own hands. The conference concluded on September 22, 2010. Three days later, Swartz set up shop at MIT.

There is not necessarily any causal connection to be found here. Swartz never announced his plans for the JSTOR documents—not publicly, at least. If he confided in friends or family members, they have kept his secret. "Maybe he was downloading them because he'd figured out a way to do it and he was going to wait to see what to do next," his friend Ben Wikler would later suggest. "Maybe he did it so he didn't have to have an Internet connection to read whatever journal he wanted."[35] Feel free to examine the evidence and draw your own conclusions—the federal government certainly did.

Whatever Swartz's intentions, the similarities between the JSTOR scrape and the PACER project are manifold. In both cases, Swartz used computer scripts to rapidly drain useful databases that weren't immediately accessible to the public, prioritizing speed and efficiency over strict compliance with those databases' terms and conditions of use. In both cases, the service providers initially framed the activity as a possible crime; that frame dictated their subsequent actions. In both cases, Swartz could have escaped notice if he had proceeded more slowly and acted within the databases' terms of use. In both cases he sacrificed caution for celerity, and in both cases he paid the price.

BY the time he started his JSTOR operation, Aaron Swartz had been living in Cambridge, Massachusetts, for more than two years. *"Cambridge is the only place that's ever felt like home,"* he wrote on his blog upon his departure from San Francisco in 2008.[36] "Surrounded by Harvard and MIT and Tufts and BC and BU and on and on it's a city of thinking and of books, of quiet contemplation and peaceful concentration. And it has actual weather, with real snow and seasons and everything, not this time-stands-still sun that San Francisco insists upon."[37]

In Cambridge, he began working remotely for the Progressive Change Campaign Committee (PCCC), a political action group he had cofounded. "We had no money and no members and not much of a plan for how to get them," Swartz wrote, but they eventually figured it out.[38] The group found its niche by executing a series of theatrical stunts that drew media attention and attracted new members. Annoyed with the omnipresence of the boisterous television host Jim Cramer, who for all his professed financial acumen had failed to foresee the collapse of the housing market in 2008, Swartz and the PCCC launched an online petition asking CNBC, Cramer's network, to hire someone—anyone—who hadn't been wrong about the subprime-mortgage crisis. "We spread the word to friends and bloggers and before we knew it we had nearly 20,000 signatures—20,000 new members," Swartz recalled.[39] The group continued to grow from there.

Swartz worked out of a shambolic old building called the Democracy Center, near Harvard Square, which had become a hub for political activists with grand ambitions and limited budgets. The adjacent office belonged to the political organizer and former *Onion* contributor Ben Wikler. Wikler was employed by a global activism group called Avaaz. The two became friends.

"I think we first had lunch July twenty-eighth of 2009. I don't remember for certain, but my big recollection is that Aaron ordered a huge plate of french fries," Wikler said. They bonded over their shared admiration for Robert Caro's book *The Power Broker*, about the urban planner Robert Moses. Swartz and Wikler soon realized they had other things in common, too. "He was game for almost anything. He was someone you didn't have to plan dinner with— you'd just say, let's go to dinner, and you'd go. We'd go to movies and talks and parties of hacker people or political fund-raisers," Wikler remembered. "We had a mutual man-crush because I knew all these people in online activism and organizing, and Aaron knew all the tech bloggers. So we were each other's tickets into the other's world."

Swartz's connections to the Cambridge tech community dated back a decade. His father, Robert Swartz, had worked as a consultant at the MIT Media Lab since 2000, providing advice on patent issues. Swartz's blog is replete with stories of him as a teenager and young adult visiting his father's office in the Media Lab. After Swartz had been accepted to the Summer Founders Program in 2005, his father had even encouraged him to live at MIT during the summer—advice he eventually took. "You could set up in the conference room," Robert Swartz suggested, not wholly in jest. "There's a shower just down the hall and coffee every morning. It'd probably be a month before [the boss] finds out."[40]

Although Aaron Swartz was never formally enrolled in or employed by MIT, he was nevertheless a member of the broader community there. Officials recalled that Swartz had been "a member of MIT's Free Culture Group, a regular visitor at MIT's Student Information Processing Board (SIPB), and an active participant in the annual MIT International Puzzle Mystery Hunt Competition."[41] Mystery Hunt is a puzzle-solving contest that is half scavenger hunt, half Mensa entrance exam. The annual event attracts participants from around the world, many of them grown adults unaffiliated with MIT. Teams spend the weekend of the hunt running around the MIT campus solving a series of difficult puzzles, occasionally sneaking into rooms and campus locations that are technically off-limits.

In Cambridge, Swartz started to treat his own life as a puzzle to be solved. He designed various lifestyle experiments to optimize his efficiency and happiness. He dabbled in creative sleep schedules, then abandoned the experiment when he found it only made him tired. In the spring of 2009, he spent a month away from computers and the Internet for the first time in his adult life. His laptop had become "a beckoning world of IMs to friends, brain-gelatinizing television shows, and an endless pile of emails to answer. It's like a constant stream of depression," he wrote. "I want to be human again. Even if that means isolating myself from the rest of you humans."[42]

He spent June offline, an experience he later described as revelatory. "I am not happy. I used to think of myself as just an unhappy person: a misanthrope, prone to mood swings and eating binges, who spends his days moping around the house in his pajamas, too shy and sad to step outside. But that's not how I was offline," Swartz wrote, recounting how he had come to enjoy simple human pleasures such as shaving and exercising in the absence of perpetual connectivity.[43] "Normal days weren't painful anymore. I didn't spend them filled with worry, like before. Offline, I felt solid and composed. Online, I feel like my brain wants to run off in a million different directions, even when I try to point it forward." Swartz vowed to find ways to sustain this serenity and thenceforth tried to make his apartment a computer-free zone. But he would never be able to avoid the Internet entirely.

In 2010, Swartz was named a fellow at the Edmond J. Safra Center for Ethics at Harvard University.[44] Lawrence Lessig, who had also returned to Cambridge from Palo Alto, brought him aboard. Lessig was supervising a Safra Center program that examined institutional corruption and its effect on public life. The fellowship was well suited for Swartz, who had spent so much of his life fixated on institutional and personal ethics. Individual ethicality had obsessed Swartz for years, and as he aged, it became perhaps his chief concern.

"It seems impossible to be moral. Not only does everything I do cause great harm, but so does everything I don't do. Standard accounts of morality assume that it's difficult, but attainable: don't lie, don't cheat, don't steal. But it seems like living a moral life isn't even possible," Swartz declared in August 2009.[45] The next month, he extrapolated from this line of thought:

> The conclusion is inescapable: we must live our lives to promote the most overall good. And that would seem to mean helping those most in want—the world's poorest people.
>
> Our rule demands one do everything they can to help the poorest—not just spending one's wealth and selling one's possessions, but breaking the law if that will help. I have friends who, to save

money, break into buildings on the MIT campus to steal food and drink and naps and showers. They use the money they save to promote the public good. It seems like these criminals, not the average workaday law-abiding citizen, should be our moral exemplars.[46]

This section ignited a debate in the comments section of Swartz's blog. Readers chided Swartz for sanctioning the theft of services from MIT. The next day, in a blog post titled "Honest Theft," Swartz defended his position: "There's the obvious argument that by taking these things without paying, they're actually passing on their costs to the rest of the MIT community." But perhaps that wasn't as bad as it seemed, since "MIT receives enormous sums from the wealthy and powerful, more than they know how to spend."

Other readers argued that the freeloaders' actions just forced MIT to spend more money on security. "I don't see how that's true unless the students get caught," Swartz responded. "Even if they did, MIT has a notoriously relaxed security policy, so they likely wouldn't get in too much trouble and MIT probably wouldn't do anything to up their security."[47] Swartz had good reason to think this way. MIT was the birthplace of the hacker ethic. The university tacitly encourages the pranks and exploits of its students; stories abound of clever undergraduates breaking into classrooms, crawling through air ducts, or otherwise evading security measures for various esoteric and delightful reasons, and these antics have been cataloged in museum exhibits and coffee-table books. By officially celebrating these pranks, MIT sends the message that it is an open society, a place where students are encouraged to pursue all sorts of creative projects, even ones that break the rules.

MIT's public reputation for openness extends to the wider world. The front doors to its main building on Massachusetts Avenue, the imposing Building 7, are always unlocked. For years, local drama groups conducted impromptu rehearsals in vacant MIT classrooms. In 2010, any stranger could show up to MIT, unfold a laptop computer, connect to its wireless Internet network, and retain the connection for a full two weeks; guests were even allowed to access MIT's library resources.

As a Safra Center fellow, Swartz had access to JSTOR via Harvard's library. So why did he choose to deploy his crawler at MIT, a school with which he was not formally affiliated? One possible reason is that computer-aided bulk-downloading violated JSTOR's stated terms of service, and, for that reason, Swartz may have preferred to remain anonymous. MIT might even have seemed to him like the sort of place that would be unbothered by, and possibly encourage, his actions. But Swartz would soon realize that MIT's public image did not directly align with reality.

THE Massachusetts Institute of Technology is not configured to address the sorts of philosophical and ethical questions that one might expect a great university to address, in part because it is not and has never been a university. It is, rather, a technical institute. (This distinction might seem like a minor semantic point, but our self-descriptions set implicit boundaries that we are often loath to cross.) Ever since its establishment in 1861, the institute has trained engineers and pursued practical applications for existing and emerging technologies. A center for applied thinking and science, it specializes in the practical rather than the philosophical. It is, as the scholar Joss Winn once put it, "*the* model capitalist university."[48]

In 1919, three years after the school relocated from central Boston to its current location on the Charles River in Cambridge, MIT president Richard Maclaurin advocated a "Technology Plan" that would help the institute foster closer ties with industry. Institute benefactor and camera magnate George Eastman, whose generous donations helped underwrite the school's relocation, had assured Maclaurin that MIT would "rise to a position of transcendent usefulness," and the Technology Plan would speed its ascent.[49]

On January 10, 1920, addressing a room of MIT alumni, Maclaurin said that the plan would allow MIT to "give industrial corporations the information that they want regarding men and scientific processes that are applicable to their industry. A mere school might not be able to do this, but an institution conceived so broadly as Technology [MIT] is well adapted for this great end."[50] Maclaurin died suddenly five days later, but the Technology Plan survived in the form of the new Division of Industrial Cooperation and Research, which was charged with marketing the school's "scientific and industrial experience and creative aptitude" to companies willing and able to purchase such things.[51] The program was financially successful, and the lessons MIT learned from its administration put the school in a position to acquire and manage millions of dollars' worth of government contracts after the United States entered World War II.

Former MIT engineering dean Vannevar Bush served as President Franklin Roosevelt's science adviser during the war and directed large amounts of money toward university laboratories in an effort to develop technologies that could aid the war effort. As Bush later noted, "World War II was the first war in human history to be affected decisively by weapons unknown at the outbreak of hostilities," which "demanded a closer linkage among military men, scientists, and industrialists than had ever before been required."[52] That linkage was particularly strong at MIT, where researchers at the school's Radiation Laboratory developed microwave radar systems for the US military and "practically every member of the MIT Physics Department was involved in some form of war work," as the department itself has stated.[53] Academic science helped the Allies win the war—and the war helped the Allies win over academic science.

In an essay for the anthology *Big Science*, S. S. Schweber wrote, "By participating massively in the war effort the Institute had transformed itself. By 1943 MIT was overseeing some $25,000,000 of government-supported contracts, whereas its total budget in 1939 was only about $3,000,000."[54] After the war, Bush and others argued for the continuance of these financial ties and recommended the creation of a federal body that would supervise and direct this conjoinment. Government support would revolutionize American academic scientific research, argued Bush in his report *Science—The Endless Frontier*, germinating a new golden age of pure science and, eventually, "productive advance." Measures would be taken to ensure the separation of scientist and state and encourage a research environment that was "relatively free from the adverse pressure of convention, prejudice, or commercial necessity."[55] Federal funding would, in fact, *save* the scientist from the indignities of industrial research. The American people would benefit from the partnership, and eventually so would the world.

Bush's benign vision for the future of academic science in the United States translated to a boon for MIT and other domestic research universities. Expansive federal support for scientific research made it easier for scientists to fund expensive experiments that had no immediate practical applications. Like privately held companies that decide to go public to fund growth and expansion, universities reaped immediate benefits from government partnerships—more money meant more hires, new facilities, and increased prestige—but they also ceded some control over their institutional priorities. In his great book *The Cold War and American Science*, the Johns Hopkins professor Stuart W. Leslie observed that the long-term effects of military-funded university research cannot be measured in merely economic terms, but must also be measured "in terms of our scientific community's diminished capacity to comprehend and manipulate the world for other than military ends."[56]

The same proposition holds true for the effect that corporate funding has had on academic science. In his seminal 1942 essay, "The Normative Structure of Science," the American sociologist Robert K. Merton famously listed the fundamental tenets of what he deemed the "scientific ethos"—the scientist's code of honor, so to speak—philosophical norms that governed inquiry and activity in all respectable laboratories, and without which no credible research could proceed. These four norms are universalism, communism, disinterestedness, and organized skepticism.[57]

Although the word has many political associations, Merton's term *communism* simply meant that the products of scientific research belong to the community, and that scientists are rewarded for their discoveries not by money, but by "recognition and esteem." Isaac Newton was the first to describe his second law of motion, for instance, but that fact does not entitle his descendants to royalties whenever an object comes to rest and stays at rest. Scientific discoveries "constitute a common heritage in which the equity of the individual producer is severely limited," wrote Merton. "Property rights in science are whittled down to a bare minimum by the rationale of the scientific ethic."[58]

Merton further noted the absolute incompatibility of the communal scientific ethos with "the definition of technology as 'private property' in a capitalistic economy."[59] As capitalism and academic science continued to coalesce in the twentieth century, this communal ethos was put at risk.

Today, MIT's own website proudly announces that it "ranks first in industry-financed research and development and development expenditures among all universities and colleges without a medical school."[60] In her fascinating 2002 dissertation, "Flux and Flexibility," the MIT doctoral student Sachi Hatakenaka traced the school's modern-day corporate partnerships and their effect on institutional structure and priorities. Though MIT has long been a cheerful collaborator with industry, the practice has expanded over the last forty years; Hatakenaka reported that industrial research funding at MIT jumped from $1,994,000 per year in 1970 to $74,405,000 per year in 1999.[61]

In the 1980s, conscious of increased competition for federal science grants and fearful that the inevitable end of the Cold War would prompt a decline in federal research funding, many American universities sought a broader range of industry partnerships.[62] Around this time, Hatakenaka noted, MIT began to encourage corporate sponsorship of its research laboratories. These alliances were attractive to corporations, as well, many of which had slashed their internal research and development budgets in the midst of economic recession. Hatakenaka wrote that consulting with industry became "something that almost all [MIT] engineering faculty members are expected to undertake, partly to supplement their own income, but more importantly, to keep up with practice."[63]

In her dissertation, Hatakenaka carefully outlined all of the boundaries that MIT has erected to ensure both that corporate sponsors have no direct control over specific research projects and that MIT researchers can maintain their scholarly independence. But the institute's increasing reliance on corporate contributions perforce affects the administration's attitude toward the free market, and toward anything that might jeopardize its profitable partnerships.

The federal support that Vannevar Bush believed would free academic scientists from the need to collaborate with industry has ended up pushing them more firmly into industry's embrace. Initially, the federal government retained title to all of the scientific research that it funded. For example, if the government gave a university lab a grant to study computing, and the laboratory used that grant money to develop a new type of microprocessor, then the government owned the rights to that microprocessor and could license those rights to private industry. This setup changed in 1980, when President Jimmy Carter signed into law the Bayh-Dole Act, which effectively privatized the fruits of publicly funded research.

Bayh-Dole was meant to address the perceived technology gap between Japan and the United States around the time of its passage, and reduce the gap between when a useful technology was developed and when it was brought to market. It decreed that, henceforth, domestic universities would retain the rights to the results of federally funded research and could patent those inventions, license them to industry, and reap the resultant profits.

After Bayh-Dole became law, universities began to establish what they referred to as "technology transfer offices": administrative divisions that existed to facilitate patent licensing and to liaise between academic researchers and corporate customers. The act allowed the fruits of academic research to be harvested and sold with unprecedented ease and rapidity, and the ensuing licensing fees made many universities wealthy. When the sale or rental of intellectual property becomes a university profit center, then research outcomes will inevitably become a proprietary concern. "Far from being independent watchdogs capable of dispassionate inquiry," wrote Jennifer Washburn in her sobering book *University, Inc.*, "universities are increasingly joined at the hip to the very market forces the public has entrusted them to check, creating problems that extend far beyond the research lab."[64]

"Free access to scientific pursuits is a functional imperative" for scientists, wrote Robert K. Merton in 1942.[65] The hacker ethic was, in a sense, a critique of applied, corporate science in the university, of the move from Mertonian universalism and communism toward proprietary research. But the hackers gradually left MIT, and the school's center for innovative computing shifted from the AI Lab to the Media Lab, Robert Swartz's employer, a loose affiliation of varied research groups that specialized in applied consumer technologies funded by a large array of corporate sponsors.

In 2008, for example, Bank of America committed $3–$5 million per year for five years to sponsor an MIT Media Lab research group called the Center for Future Banking. "We are bringing together the creative, multidisciplinary research of Media Lab faculty and students with the real-world business experience and deep-domain knowledge of our Bank of America colleagues—all in a highly innovative environment that promotes unconventional thinking and risk-taking," the director of the Media Lab said at the time.[66] The AI Lab hackers had hoped that their work would change the world. The Media Lab researchers just wanted to make it easier to bank there.

The unlocked campus, the student pranks, the accessible computer network: these are the public trappings of translucence, vestiges of an era when MIT *did* perhaps take it seriously, or else decoys to deflect attention from the fact that it had never really done so. In 2002, when he was still fifteen, Swartz traveled to MIT to speak about the Semantic Web. At that time, MIT's wireless Internet network did *not* allow guest access, and Swartz had to use a public computer terminal to get online. Unfortunately, the only non-password-protected terminals he could locate were behind a locked door. "I joked that I should crawl thru the airvent and 'liberate' the terminals, as in MIT's hacker days of yore," Swartz recalled on his blog. "A professor of physics who overheard me said 'Hackers? We don't have any of those at MIT!' "[67]

ON December 26, 2010, JSTOR realized that the MIT hacker had returned. "Woot . . . mit scraper is back," read the subject line of an internal e-mail speculating that the hackers were working out of the Dorrance Building on MIT's campus.[68] "87 GB of PDFs this time, that's no small feat, requires Organization," one JSTOR employee wrote. "The script itself isn't very smart, but the activity is organized and on purpose."[69]

The harvesting sessions had ceased only because Swartz had been out of town for a couple of months. In mid-October of 2010, Swartz traveled to Urbana, Illinois, the hometown of Michael Hart, to speak at Reflections | Projections, a conference hosted by computing students at the University of Illinois. (There is no evidence that Swartz and Hart ever met or corresponded.) Swartz's topic was "The Social Responsibility of Computer Science," and he argued that computer programmers have an ethical responsibility to advance the public welfare. He spoke about utilitarianism, and the coder's special ability to write simple programs that could automate and speed tedious, mundane activities, and complete countless tasks in the time it would ordinarily take to complete just one. "Now, as programmers, we have sort of special abilities. We almost have a magic power," Swartz said. "But with great power comes great responsibility, and we need to think about the good that we can do with this magical ability. We need to think about, from a utilitarian perspective, what's the greatest good we can achieve in the world at small cost to ourselves?"[70]

In early November 2010, Wikler and Swartz went to Washington, DC, to volunteer for the Democratic National Committee in the days preceding that year's midterm elections. Swartz was assigned to work under Taren Stinebrickner-Kauffman, a political activist who served as project manager for a telephone-outreach tool that helped volunteers contact voters in key states and districts. "Nobody really knew Aaron, so he sort of got plopped onto my team and, needless to say, was very helpful," Stinebrickner-Kauffman remembered. "It's hard to imagine a better last-minute volunteer to come in when you're working on a political technology project."[71]

When Swartz wasn't working on robocalls and performing other menial tasks for the DNC, he was observing how campaign technology worked and thinking about ways to make it work better. "We were crashing at a friend's place, talking until five in the morning afterwards, talking about how technology could do something vastly more powerful and politically impactful," Wikler recalled.[72] The election ended, and Wikler tried to persuade Swartz to remain in the capital for a few days to attend Stinebrickner-Kauffman's birthday party. Swartz declined and returned to Cambridge to revisit his own powerful and politically impactful utilitarian project.

In November of 2010, he found a wiring and telephony closet in the basement of the Dorrance Building—also known as Building 16—jacked his laptop directly into the campus network, and resumed his downloading. Swartz had refined his tactics: the script no longer triggered any of JSTOR's download thresholds. (The revolution will, after all, be A/B tested.) The downloads weren't detected until late December.

"I am starting to feel like they [MIT] need to get a hold of this situation and right away or we need to offer to send them some help (read FBI)," an aggravated JSTOR staffer wrote on December 26.[73] At the time, MIT's libraries had closed, and the school's librarians were on budget furlough until January and unable to do any work in the meantime. Thus Ellen Finnie Duranceau of MIT didn't receive any of JSTOR's increasingly frantic e-mails until January 3. On January 4, she noted that MIT was unlikely to be able to identify the culprit. "I wish I could say otherwise," she wrote, "because I realize that JSTOR would like more information and would like us to track the downloaded content to the source."[74]

But by the time Duranceau sent JSTOR the disappointing news, an MIT network engineer had already traced the downloads to a network switch in the Building 16 wiring closet. When he entered the closet, the engineer immediately noticed something odd. Though MIT exclusively uses light blue Internet cables, an off-white cable was plugged into the network switch, leading from the switch to an object concealed under a cardboard box. Lifting the box, the engineer discovered Swartz's laptop. He called one of his colleagues. Then he called the MIT police.[75]

The MIT police, concluding that the investigation required specialized skills that they did not themselves possess, promptly called a Cambridge police detective named Joseph Murphy, who belonged to a regional computer-crime task force. Murphy drove to the scene, accompanied by two other task-force members: a Boston police officer named Tim Laham and a Secret Service agent named Michael Pickett. They arrived at the Dorrance Building's basement around 11:00 a.m., and soon the wiring closet was outfitted with a motion-activated camera connected to the campus security network and devices that would log the laptop's download activity and alert MIT officials if the computer was removed from the network.[76] At 3:26 p.m. on January 4, the camera captured footage of Swartz entering the closet to check on the laptop and swap out hard drives. The police had a face. Now they needed a name.

"I've just had an update," Duranceau wrote to her JSTOR contact on January 5.[77] "The investigation has moved beyond MIT and is now being handled by law enforcement, including federal law enforcement." Duranceau was referring to Agent Pickett, who would subsequently become very active in the investigation. "If you have the time, I would appreciate if you would take a look at a new development that came to our attention yesterday," Pickett wrote in an e-mail to the US Attorney's Office in Boston on January 5, 2011, noting that the task force had discovered a laptop at MIT downloading valuable technical journals, and that the laptop matched the description of one that had been stolen from the MIT Student Center days before. (It wasn't actually the same laptop.) "I would like to get your opinion on what offenses the suspect could be charged with in this case and what evidence would best support prosecution."[78] ("Thanks for all of the detail—very helpful" was the response. "Steve and I will discuss this and get back to you by the end of the day.")

The next morning, Steve—that is, Assistant US Attorney Stephen Heymann, the office's resident computer-crime specialist—responded to the agent with a blank e-mail and a terse subject line: "Please call. Steve."[79] That same morning, the investigators made plans to remove Swartz's laptop from the closet, dust for fingerprints, and image its hard drive.[80] But Swartz returned before they had a chance to do so. He entered the closet shortly after noon on January 6, and this time, he covered his face with a bicycle helmet, as if he suspected that cameras had been installed. Though MIT police captain Jay Perault was watching the video feed in real time, Swartz was too quick for him: in less than two minutes, he disconnected the laptop, retrieved it, and left the closet before officers could scramble to the scene.[81] "It is gone, he just left—my guys are looking for him," an MIT officer wrote at 12:55 p.m.[82]

From there, things moved quickly. A couple hours later, MIT police captain Albert Pierce spotted someone who resembled the man in the video riding his bicycle up Massachusetts Avenue toward Central Square, away from MIT.[83] When Pierce approached, Aaron Swartz informed him that he didn't talk to strangers. Pierce showed Swartz his badge, but Swartz wasn't impressed, retorting, "MIT Police were not 'real cops.'"[84] Then Swartz dropped his bicycle and ran.

Pierce chased after him, but could not keep pace, so he followed Swartz in his car. Jay Perault joined the chase, accompanied by Michael Pickett. Swartz was two blocks from his apartment when his pursuers ran him down on Lee Street. They surrounded Swartz in a parking lot. They chased him through the cars. They caught him, handcuffed him, and brought him to the station. He initially refused to identify himself to police.[85]

He called Quinn Norton, who called Swartz's lawyer, who bailed him out of custody. "I think I saw him on the day he was arrested," Ben Wikler remembered. "He was just totally, totally freaked-out. White as a ghost." Earlier that morning, Swartz had posted on Twitter a quote by the philosopher Willard Van Orman Quine: "'Ouch' is a one-word sentence which a man may volunteer from time to time by way of laconic comment on the passing show."[86]

EVEN if Swartz hadn't broken a law, he had apparently violated JSTOR's usage policies. JSTOR allowed walk-in users with no formal university affiliation to use its services as long as they were "physically present on the Institutional Licensee's premises" while accessing the database.[87] Swartz was not physically present while his script was deployed and thus wasn't explicitly authorized to access the database. Moreover, JSTOR's terms of service stated that users were not allowed to deploy scrapers or spiders or any other computer programs that might overtax the database's servers. Users were also forbidden from downloading entire issues or volumes of journals unless they could articulate specific research purposes for doing so.[88]

Swartz was initially charged with breaking and entering, but the authorities immediately started looking for ways to charge him with more. In 1997, Stephen Heymann had written an article for the *Harvard Journal on Legislation* called "Legislating Computer Crime," in which he argued that Congress needed to take action to address the unique challenges posed by digital scofflaws: "The ability of computers to perform a task millions of times in the period that it takes a human to perform the same task only once dramatically increases the harm that a particular action can cause."[89] The skill set that Swartz extolled as a "magic power" to his audience in Urbana was seen by the government as an evil spell.

Heymann had a powerful counterspell at his disposal to fight those digital warlocks: the Computer Fraud and Abuse Act (CFAA). Congress passed the first version of the CFAA in 1984 to stop malicious nerds and undersexed teenagers from hacking into banking and government computers, as happened in the Matthew Broderick film *WarGames*. (*WarGames* was mentioned more than once in the committee hearings preceding the bill's passage.) After several rounds of amendment and revision, the law eventually established criminal penalties for anyone caught accessing a protected computer without authorization, and defined a protected computer as any machine engaged in interstate commerce or communication.

Today, any computer with an Internet connection is engaged in interstate communication; likewise, most institutional websites feature terms-of-service agreements that visitors actively or implicitly accept upon arrival. Violating these terms of service can count as accessing a protected computer without authorization. While it is unlikely that federal charges would ever be filed against a teenage boy who clicks the *Over 21* box on a beer company's website, the CFAA's imprecision gives prosecutors the latitude to hang felony charges on unlikely defendants.

In 2005, during his trip to Cambridge to pitch Paul Graham on Infogami, Swartz visited his father's office in the Media Lab. "Arguing about the merits of MIT, my dad plays the Noam Chomsky card," Swartz wrote. One thing led to another, and the two set out on an impromptu search for Chomsky's office. (They thought they found it, but they weren't entirely sure.)[90]

What Swartz may or may not have understood at the time was that MIT's employment of Noam Chomsky was not the same thing as an endorsement of his political critiques. Nor did it mean that the institute wasn't yet another of the authoritarian institutions that Chomsky dissected in his books. If you're interested in understanding power, you have to understand how power perpetuates itself, how it is wielded like a cudgel to bludgeon deviants until they surrender or shatter.

In his 1997 article, Heymann cited the CFAA as model legislation, calling it "a computer crime law at its purest." In 2011, he was looking for an opportunity to bring it to bear on Swartz. On February 16, Heymann e-mailed his JSTOR contact asking, among other things, "When you have a moment, can you give me a call to discuss loss valuation?"[91] You mess with the bull, you get the horns.

9

THE WEB IS YOURS

At the beginning of every year, Aaron Swartz would post to his blog an annotated list of the books he had read over the previous twelve months.[1] His list for 2011 included seventy books, twelve of which he identified as "so great my heart leaps at the chance to tell you about them even now."[2] The list illustrated the depth and breadth of Swartz's interests. There was *CODE: The Hidden Language of Computer Hardware and Software*, by Charles Petzold ("I never really felt like I understood the computer until I read this book"); *The Lean Startup*, by Eric Ries ("Read it with an open mind and let it challenge you, so you can start to understand how transformative it really is"); *The Pale King*, an unfinished posthumous novel by David Foster Wallace, Swartz's favorite fiction writer ("Probably less unfinished than it feels").

The list also included Franz Kafka's *The Trial*, about a man caught in the cogs of a vast judicial bureaucracy, facing charges and a system that defied logical explanation. Pertinent information is withheld from the novel's protagonist, Joseph K.; his situation grows less scrutable over time. The system in which he is trapped does not judge its subjects so much as erode them. "I read it and found it was precisely accurate—every single detail perfectly mirrored my own experience," Swartz wrote. "This isn't fiction, but documentary."

His arrest in January had been a shock; its subsequent ramifications were just as jarring. "Neither of us seemed able to believe this was serious," his girlfriend at the time, Quinn Norton, later wrote.[3] The JSTOR downloads, Swartz and his supporters rationalized, equated to withdrawing too many books from a library. Why would the government care about that? Yet Swartz was facing two state felony breaking-and-entering charges. He had been barred from the MIT campus. Federal prosecutors were starting to build a case against him. In his notebook, Swartz wrote a story about walking to breakfast in Cambridge, feeling chest pains, and being rushed by ambulance to Beth Israel Hospital. "Well, it's probably just stress," the doctor said in Swartz's story. "You're too young to be having heart problems."[4]

Soon after his arrest, Swartz tried to contact JSTOR, perhaps thinking that an apology and an explanation might make the bad thing go away. "If you know someone who works at JSTOR (or ITHAKA), please send me an email: me@aaronsw.com. Many, many thanks," read a January 10 post on Swartz's Twitter account.[5] "Do you know Kevin Guthrie?" Swartz asked Carl Malamud on February 6. (Guthrie was the president of JSTOR's parent organization, ITHAKA.) "no, but I'm a big fan of Arlo and Woody," Malamud replied.[6]

While Swartz scrambled to find a JSTOR contact, the US Attorney's Office had already worked its way inside. On January 10, Stephen Heymann held a conference call with a JSTOR official "to discuss the theft of material from JSTOR," according to the Secret Service.[7] Heymann continued to accumulate evidence against Swartz, deeming him responsible for JSTOR's ostensible loss of revenue from the articles he downloaded, as well as the expenses JSTOR had incurred in fixing the servers that Swartz had crashed. (Withdrawing too many books from a library is one thing; toppling the shelves on

your way out the door is another.) The police had seized the laptop that Swartz used to connect to the MIT network, as well as a
thumb drive containing a version of the downloading program that
Swartz had used. "We are taking our time with this investigation and
crossing all the T's etc," a Secret Service agent wrote in a January 20,
2011, memo. "Case continuing, more to follow."[8]

In 2009, the FBI had eventually dropped its investigation into
Swartz's PACER downloading spree. But unlike the court records
Swartz had siphoned from PACER, the JSTOR articles were copyrighted documents that qualified as private property. Swartz's governmental antagonists were unimpressed by his reputation as an
Internet icon and unwilling to cut him slack because of his youth
and achievements. "Looks like he is a big hacker, i googled him," was
one MIT police officer's response upon Swartz's arrest.[9] Not *Reddit
cofounder*; not *Open Library architect*; not *computer prodigy* or
applied sociologist or *Harvard affiliate* or any of the other lines on
his résumé. *A big hacker*. And a suspicious person might well read
some of Swartz's overheated free culture rhetoric and conclude that
he was a malicious one.

Ascertaining Swartz's motives would be key to the federal case.
The prosecutors knew he had downloaded the JSTOR documents,
but they didn't know why, and Swartz was determined to keep it that
way. His blog went silent for months after the arrest, as if he was
reluctant to say anything that the prosecutors might theoretically
be able to use against him. On February 1, Malamud asked Swartz
for some Department of Justice documents that Swartz had recently
acquired through a Freedom of Information Act (FOIA) request.
"i'll try to get the FOIA docs scanned at some point, but i'm trying
to lie low at the moment," Swartz replied.[10] He told almost no one
about his situation.

On February 9, 2011, the government obtained a warrant to
search Swartz's one-bedroom apartment at 950 Massachusetts Avenue in Cambridge. A Cambridge Police officer staked out the apart-

ment that evening, but found it abandoned: "The unit has no sign of life. The curtains are wide open and the lights have not flickered once," the surveilling officer noted.[11] The next morning, perhaps worried that Swartz had skipped town, the Secret Service issued an "URGENT Request" for records of any airplane flights Swartz might have booked.[12] But Swartz eventually returned home, and the Secret Service executed the warrant on the morning of February 11, seizing a modem, a hard drive, some rewritable compact discs, his phone, his notebook, and a few other items.

Swartz presented a brave face during the search, asking the police what had taken them so long to get around to searching his apartment. But the agents' presence upset him, and his bravado quickly wilted. The official report details how, midsearch, an agent "observed Swartz leave the building, walk to the street and sprint away after he reached the street."

Swartz headed for the Safra Center for Ethics at Harvard. The agents eventually followed, and they had Harvard police officers secure Swartz's office while they obtained a federal search warrant. They seized Swartz's work computer, a portable hard drive, and his controller for the video game *Rock Band* and left him there, humiliated and deprived of devices.[13] Harvard University eventually suspended Swartz's Safra Center fellowship and banned him from campus.[14] His isolation grew.

The same day that the Secret Service obtained a warrant to search his apartment, Swartz had presented three papers in a seminar at the Safra Center. The best-received among them was an "institutional ethnography" of Congress titled "How Congress Works." In it, Swartz argued that the wealthiest members of society set the congressional agenda; that "even new members of Congress are surrounded and advised by individuals whose interests are far from being aligned with those of the elected officials' constituents," as a Safra Center recap put it.[15]

Misaligned congressional priorities were more than just a matter of academic interest for Swartz. In 2010, as part of his work with the Progressive Change Campaign Committee, Aaron Swartz had met a young Rhode Island state representative named David Segal, who was running for the US Congress. Segal promised to be a progressive voice in Congress, and, moreover, an Internet-friendly one. His campaign was underfunded and at times endearingly homemade: one campaign advertisement featured several crude puppets, to emphasize that Segal would be no one's puppet in Congress. Segal lost by a wide margin in the primaries.

Soon thereafter, Swartz left the PCCC and, together with Segal, founded the advocacy organization Demand Progress: a political action group that would address a wide range of issues pertaining to civil liberties, social justice, and the Internet. The group hoped to bridge the gap between the people who used the Internet and the people who hoped to regulate it by giving Internet users new ways to speak out on their pet issues. They also hoped that this dialogue would force lawmakers to address and contend with opinions that they were not accustomed to hearing.

Their marquee tactic was the online petition, which, if Washington lobbying techniques were described as weapons, would probably be deemed the equivalent of a popgun, or a very small rock. But well-aimed rocks have been known to topple giants, and Demand Progress hoped to leverage its executives' social networks and technical prowess to obtain signatures in numbers that couldn't be ignored. Their first opportunity to test this theory occurred in November 2010, when the US Senate Judiciary Committee considered a new bill called the Combating Online Infringement and Counterfeits Act (COICA).

Two longtime intellectual property hard-liners, Senators Patrick Leahy of Vermont and Orrin Hatch of Utah, introduced the bill. COICA was designed to fight copyright infringement by targeting foreign and domestic websites that hosted pirated content: illicitly uploaded movies or music or similar files. COICA empowered the

Department of Justice to seek injunctions against these sites and effectively take them offline via a process called DNS blocking. The bill would also allow the government to seek court orders to prevent other websites from doing business with these offenders. "Protecting intellectual property is not uniquely a Democratic or Republican priority—it is a bipartisan priority," said Leahy upon the bill's introduction.[16] The bill's eight initial cosponsors, who sat on both sides of the aisle, presumably agreed.

Although COICA was framed as a weapon against "file lockers"—such as the website Megaupload, founded by an obese, gaudy man who went by the name Kim Dotcom; or alleged bootleg-film havens with names such as The Pirate Bay—many feared that it carried broader implications for the Internet at large. "It gave the government vast new powers to censor the Internet," Swartz would later say. "The attorney general could go to a court, and without proving that a website had committed any crime, or violated any law, they could get that website shut down. Taken off the Internet, so that nobody could access any piece of it."[17]

Stopping COICA became Demand Progress's first campaign. In November 2010, the group launched a petition against what it termed the "Internet Blacklist Bill," and as online petitions go, it was wildly successful. "I've worked at some of the biggest groups in the world that do online petitions," Swartz later said. "But I've never seen anything like this. Starting from literally nothing, we went to ten thousand signers, then a hundred thousand signers, then two hundred thousand, then three hundred thousand."[18]

Still, the petitions did not directly impede the bill's momentum. COICA passed the Senate Judiciary Committee by a unanimous vote and seemed destined to come up on the floor of the Senate, until Senator Ron Wyden placed a "hold" on the bill, thus delaying its progress. In March 2011, the congressional session and COICA expired. It was a small and unsatisfying victory. Swartz and Segal were sure that COICA or something similar would soon return. Like reanimated corpses in a horror movie, regressive computer laws just kept coming, unable to be stopped, only slowed down or outrun.

THE prosecutors offered Swartz a pre-indictment plea deal of a few months in prison followed by a period of supervised release. Swartz gave it serious thought, but was ultimately unwilling to spend time in prison or have a felony conviction on his record. He declined the deal. The government continued to build its case.

"Identify Associates," a Secret Service agent working the case wrote in a notebook after Swartz's arrest.[19] In March 2011, the US attorneys subpoenaed Quinn Norton—Swartz's most intimate associate—in hopes that she might be able to elucidate his intentions. The prosecutors could reasonably assume that Norton had some foreknowledge of Swartz's actions: the two had been dating for years. But Norton, who was in San Francisco at the time the Secret Service found her, claimed that the details of the JSTOR hack had been just another of Swartz's many secrets.

"I was terribly angry at the very thing I said would keep me safe: my ignorance of his JSTOR activities," she wrote. "Why hadn't he told me anything? Why had he let me get sideswiped by all of this?"[20] In her initial interview with the Secret Service agents who served her with the subpoena, Norton averred that she had not been in recent contact with Swartz, but she had heard that Swartz had been banned from the MIT campus, and that he had been upset that the ban would prevent him from participating in the 2011 Mystery Hunt. She claimed that she didn't even know what JSTOR was.[21]

Norton was summoned to appear before a grand jury, but the prosecutors asked if she would be willing to talk with them beforehand. They offered her a "Queen for a Day" deal, in which whatever she said in conversation with the prosecutors could not be used against her in court. Norton, a journalist who covered Internet culture and computer hacking, was worried that if she didn't cooperate prosecutors might try to seize her computers, which contained material that could be used to identify some of her journalistic sources. Swartz did not want Norton to meet with Heymann. "Aaron told me his lawyer was angry too, that I was being an idiot," Norton wrote. "He wondered, loudly, whose side I was on."[22] She reluctantly met with the prosecutors on April 13, 2011.

The government's notes from the "proffer" session indicate that Norton walked prosecutors through Swartz's life and work, and recounted the day of his arrest. She told them that Swartz was solitary and prone to depression. She mentioned his work with Open Library and Demand Progress. She explained that academics despised the academic publishing system, and that Swartz didn't like it much either.[23] Norton wrote about the meeting two years later in a piece for the *Atlantic*, noting that she was on the prescription painkiller Vicodin during the interview and intimidated by the courthouse setting. She portrayed the prosecutors as aggressive and eager to exploit her woozy state and relative unfamiliarity with the criminal justice system. The prosecutors didn't believe that Swartz had kept the JSTOR downloads a secret; they pressured her to give them something, anything that might explain or contextualize Swartz's actions. Norton mentioned the "Guerilla Open Access Manifesto," which the prosecutors had not yet seen.

The manifesto was a public document, then over two years old. "We need to take information, wherever it is stored, make our copies and share them with the world," Swartz had written. "We need to download scientific journals and upload them to file sharing networks." To the prosecutors, the fact that Swartz had done the former implied that he had also been planning to do the latter. This thesis would define their case.

Almost immediately, Norton realized that volunteering this information had been a mistake. She had provided the prosecution with a motive. "I opened up a new front for their cruelty," Norton lamented in the *Atlantic*. "Four months into the investigation, they had finally found their reason to do it. The manifesto, the prosecutors claimed, showed Aaron's intent to distribute the JSTOR documents widely. And I had told them about it."[24]

Swartz was livid. "Aaron told me that Steve [Heymann] had been viciously gleeful to [Swartz's lawyer Andrew Good] about the manifesto," Norton wrote, "that he'd said Aaron would never get as good a deal as he'd turned down now that they had that bit of evidence."[25] Divulging the existence of the manifesto felt like a betrayal to Swartz, and it drove a wedge between him and Norton. Their relationship started to come apart.

The case came together at an agonizingly slow rate. By mid-April 2011, three months after Swartz's arrest, no federal charges had yet been filed. Though Swartz did his best to maintain a sense of normalcy—he started blogging again, though not as often as before; he assisted with the relaunch of a magazine called the *Baffler*—other aspects of his life were on hold. His lease on the unit at 950 Massachusetts Avenue expired at the end of May, and he asked to renew on a month-to-month basis, unwilling to commit to an extended stint in Cambridge.

Swartz also began to grow paranoid and worried that he would be arrested and taken away with no warning. That May, still in prosecutorial limbo, Swartz called a *Wired* editor named Ryan Singel and tipped him that "the feds might come knocking" on his door soon, in response to an excessive downloading incident about which Swartz was unwilling to elaborate. "Aaron, whom I'd written about before, was being careful—which meant cagey, evasive, and awkward," Singel wrote later. "'I hope this doesn't happen,' he added, referring to the raid."[26]

The prosecutors were in no rush to deliver an indictment. That spring, they attempted to determine whether Swartz had used his Harvard University computer in committing a crime. (They suspected that Swartz had used his Harvard computer to remotely check the progress of the JSTOR downloads.) They also investigated the possibility that Swartz had siphoned material from *other* proprietary databases, too. ("! !—Similar Activities—," a Secret Service agent scribbled in a notebook.)[27] And they remained fascinated by the "Guerilla Open Access Manifesto."

On May 9, 2011, a Free Software Foundation employee named Joshua Gay, an acquaintance of Swartz's, was summoned to appear before a grand jury to answer questions about "Geurilla [*sic*] Open Access." (During this period, the Secret Service consistently botched the spelling of *guerilla*.) Gay had created a Guerilla Open Access Facebook page, and the prosecutors wanted to determine what, if anything, he knew about Swartz's plans.[28] Free software, open access: ideas that had been advanced as matters of justice now counted as potential crimes.

Three days later, on May 12, Senator Patrick Leahy introduced a new bill called the Preventing Real Online Threats to Economic Creativity and Theft of Intellectual Property Act of 2011, otherwise known as the PROTECT IP Act, or PIPA. The bill, which featured eleven initial cosponsors, was a cousin to COICA. Whereas COICA had focused on both foreign and domestic websites that illegally hosted copyrighted content, PIPA narrowed its ambit to target foreign offenders. The new bill added a clause that would force search engines such as Google to remove offending sites from their search results; it also allowed individual copyright holders to pursue injunctions against offending websites.

Alongside other groups such as the Electronic Frontier Foundation and Public Knowledge, Swartz and Demand Progress tried to galvanize opposition to PIPA. "COICA was bad, but PROTECT IP is worse," cried the announcement on the Demand Progress blog. "We've been fighting these bills since their introduction, and now we need your help more than ever. Check out the full text of PROTECT IP bill and then SIGN OUR PETITION and help us keep the internet neutral and open!"[29] But THEIR PETITION, along with the efforts of their allies, wouldn't be enough to stop the bill.

PIPA unanimously passed through the Senate Judiciary Committee on May 26, 2011, without even so much as a hearing. "This legislation will provide law enforcement and rights holders with an increased ability to protect American intellectual property," Senator Leahy remarked upon the bill's approval. "This will benefit American consumers, American businesses, and American jobs."[30]

To Swartz and his allies, the economic benefits of PIPA touted by Senator Leahy were far outweighed by the bill's deleterious implications for free speech online. Senator Wyden placed a hold on PIPA the day that the bill moved to the full Senate.[31] But it was early in the legislative session, and the hold wouldn't last forever. With nearly a third of the Senate eventually attached as cosponsors, the bill seemed destined to become law. "Demand Progress was substantially under-resourced, and certainly wouldn't win this fight on its own, or as party to the small coalition that was responsible for organizing the bulk of the anti-COICA and PIPA work to date," David Segal later wrote. "We'd have to fend off the bill's backers long enough to build a more robust coalition, or for somebody to intervene from the heavens."[32]

While divine intervention did not appear imminent for Aaron Swartz, his lawyers hoped at least to persuade JSTOR to say a few words on their client's behalf. Several months after Swartz had been arrested, JSTOR officials' enthusiasm for the case had waned. A Secret Service agent noted that "JStor upset about 1) Desperately Want Back What was taken; 2) Worried about be bad guy."[33] If Swartz could fulfill the former priority, then maybe JSTOR would be willing to discuss ways to avoid the latter.

In June 2011, as part of a civil settlement negotiated by his attorneys, Swartz turned over to the US Attorney's Office the JSTOR articles he had downloaded, delivering to them a disk that he swore contained the only copies of the documents. He also agreed to pay JSTOR $26,500 for damages and attorney fees.[34] In exchange, JSTOR's attorneys called the US Attorney's Office to say that the organization had no interest in seeing Swartz prosecuted.[35]

MIT was not as cooperative. The institute sought no restitution from Swartz and was uninterested in pursuing a civil settlement. MIT had remained publicly silent since the arrest, and when a JSTOR official contacted MIT Libraries in mid-June to discuss the possibility of a joint statement in the event of Swartz's indictment, the libraries' representative replied that the school's attorneys "believe in general that the less MIT says, the better. We can't really discuss the details of the ongoing criminal investigation and possible indictment, nor do we want to interfere with the processes and duties of the USAO."[36]

Swartz's advocates worked hard to get MIT to reconsider. They hoped that a sympathetic public statement from the school might dissuade the US attorneys from bringing a case. At the least, such a statement would hamper the prosecutors' efforts to argue in court that Swartz had caused grave harm to any of the parties involved.

On June 13, 2011, Robert Swartz asked Joi Ito, the incoming director of the MIT Media Lab, if he would be willing to intervene on Aaron's behalf with the MIT administration. Ito obliged. "I wonder if there is any way that MIT might consider this a 'family matter' and consider helping to try to limit the extent of the punishment and at least prevent Aaron from going to prison on a felony charge," Ito wrote to the school's Office of the General Counsel. "Obviously it was a stupid thing to do, but the weight of the possible sentence seems quite harsh in my personal opinion."[37]

But MIT had already decided to remain institutionally neutral. Swartz was not formally affiliated with MIT. Moreover, he had been careless with the institute's computing resources and institutional relationships. The school saw no reason to speak out on his behalf.[38] Despite many subsequent entreaties from Robert Swartz—despite his best efforts to get the institute to see the "human side of the story"[39]—MIT would not waver from this stance.

On June 16, 2011, Quinn Norton testified before the grand jury in Boston.[40] Her earlier deference to federal prosecutors had turned into open antagonism. "When I admitted I wasn't surprised at all at [Swartz's arrest and the subsequent investigation], they asked me why," she wrote later. "I told them because there was a trend towards overreaching police action plaguing the tech community and seeking to criminalize normal computer use and research."[41] Despite her defiant testimony, an indictment was on its way.

As his relationship with Norton disintegrated, Swartz began to date Taren Stinebrickner-Kauffman, a political activist who lived in Washington, DC. She was tall and pale, with a toothy grin and a technical mind. "I grew up as the kind of kid who skipped the prom to go to an international science fair," she wrote of herself on her blog.[42] Like Swartz, she had engineered an early exit from high school.

Swartz met Stinebrickner-Kauffman in November 2010, when he and Ben Wikler had volunteered at the Democratic National Committee during the midterm elections. "As soon as the polls closed in Hawaii, and everyone was beginning to wind down, I turned to Ben and I asked him if Aaron was flirting with me," she remembered. "And he said he didn't know, but would it be a good thing if he was? And I said, 'Yeah, probably.' "[43]

Over the next few months, their paths crossed several times, and a mutual attraction developed. In February, they had both traveled to Madison, Wisconsin, to join with public-sector unions who were protesting the regressive policies of the state's new governor, Scott Walker. "I was trying very hard to flirt with him without a whole lot of success," Stinebrickner-Kauffman recalled, "although in retrospect that was right after he had been arrested and stuff, so he might have been a little preoccupied."[44] That June, Stinebrickner-Kauffman traveled to Boston for work and flat-out asked Swartz if he'd like to go on a date. He said yes.

"We went to a Chinese restaurant on Mass Ave. [in Cambridge], somewhere near Central Square. He told me he was really busy and had a lot of things going on, which I didn't really know what they were," Stinebrickner-Kauffman said. The date, which went well, ended on a mysterious note: "He said that he didn't think he'd be able to see me for, like, six weeks, that his schedule was so packed. Which I was, like, a little disgruntled about."[45] But Swartz rearranged his schedule to make room for romance: that weekend, he flew to Washington, DC, to see Stinebrickner-Kauffman.

As June turned to July, Swartz had stepped down from his position as executive director of Demand Progress, in anticipation of being charged with a crime. He had officially ended his long-term relationship with Quinn Norton. He had revised his will and made a bequest to a charity called GiveWell. "He explained that he wanted to use his money to accomplish as much good as possible, and that as long as he was alive, this meant funding projects of his own," GiveWell's cofounder, Holden Karnofsky, later wrote, "but that if something unexpected happened, he wanted the money to go to the next-best option."[46]

Swartz had been circumspect with Stinebrickner-Kauffman, referring to his legal troubles only as "the bad thing." Early in their relationship, Stinebrickner-Kauffman and Swartz were walking down Massachusetts Avenue in Cambridge toward Boston, near the MIT campus. "I had to go to the bathroom, and for some reason he waited for me out on the sidewalk as I ran into the cafeteria building instead of coming in with me," Stinebrickner-Kauffman later wrote.[47] "I didn't think much of it at the time." She didn't know that he had been barred from campus.

But the time for circumspection was coming to an end. On July 18, Swartz called Stinebrickner-Kauffman and told her that something big would be in the news the next day. Stinebrickner-Kauffman later recounted the conversation on her blog: "He said, 'I'm going to be indicted for downloading too many academic journal articles, and they want to make an example out of me.' And I said, 'That doesn't sound like a very big deal.' He paused for a second and thought about it and said, 'Yeah, I guess it's not like anybody has cancer.' "[48]

Even seven months into his ordeal, it was still occasionally difficult to believe that it was worth taking seriously.

THE next day, the US Attorney's Office in Boston announced that Swartz had been indicted in federal court on four felony charges. Swartz reported to the John Joseph Moakley United States Courthouse on the Boston waterfront, where he was arrested by US marshals, then photographed and fingerprinted and freed on a $100,000 bond. In his booking photograph, he wore a gray, collared shirt and a pursed, inscrutable smile—half nausea, half I'll-never-tell.[49]

Swartz was accused of wire fraud—for downloading the JSTOR articles "by means of material false and fraudulent pretenses"; computer fraud—for accessing protected computers on MIT and JSTOR's networks "without authorization and in excess of authorized access"; unlawfully obtaining information from a protected computer—for accessing materials with a value in excess of $5,000; and recklessly damaging a protected computer—for causing loss to MIT and JSTOR "aggregating at least $5,000 in value and damage affecting at least 10 protected computers." The latter three charges had been brought under the Computer Fraud and Abuse Act.

The indictment recounted the extent of Swartz's JSTOR activities in great detail. During November and December 2010 alone, Swartz had downloaded from JSTOR "more than one hundred times the number of downloads during the same period by all the legitimate MIT JSTOR users combined." He had used a program called keepgrabbing.py to siphon the articles. After retrieving his computer from the switching closet on January 6, Swartz had taken it to the MIT student center and reconnected to the network wirelessly, with a new IP address. The indictment flatly asserted that "Swartz intended to distribute a significant portion of JSTOR's archive of digitized journal articles through one or more file-sharing sites." If convicted on all charges, Swartz faced a maximum penalty of thirty-five years in prison and a $1 million fine. Nobody believed he would ever get anything close to a maximum sentence, but the specter of those thirty-five years hovered over him always, like a show of power, a symbol of fear.

"Federal Government Indicts Former Demand Progress Executive Director for Downloading Too Many Journal Articles," read the headline of the post on Demand Progress's blog that afternoon. In the blog post, David Segal compared the charges against Swartz to "trying to put someone in jail for allegedly checking too many books out of the library."[50]

Another post encouraged readers to sign an online petition confirming that they "stand with Aaron Swartz and his lifetime of work on ethics in government and academics."[51] Within hours, fifteen thousand people had signed the petition, and Demand Progress had published another post expanding its criticisms of the case against Swartz: "Downloading data made available on a network is not 'stealing.' And he made copies of documents. He did not 'take' them. JSTOR still had the documents."[52]

By the end of the week, forty-five thousand people had signed the petition, and the indictment had been mentioned in the *New York Times*, the *American Prospect*, the *Boston Globe*, and other media outlets. Swartz was "an Internet folk hero," said the *Times*.[53] The charges against him amounted to using "a sledgehammer to drive a thumb tack," a local copyright lawyer told the *Boston Globe*.[54] "As the case proceeds, we remain hopeful that Aaron will be cleared of any wrongdoing," said Segal, "and as has been proved over the last 24 hours, the more people learn about [the] case, the more sympathetic they become to Aaron's cause."[55]

While Swartz may have won the sympathies of the public and the press, the US Attorney's Office in Boston was unmoved by the display. After delivering the indictment, Heymann had refrained from sending the police to arrest Swartz, instead allowing him to come to the courthouse under his own power. Swartz repaid that kindness by starting what Heymann later characterized as a "wild Internet campaign."[56] As with COICA and PIPA, Swartz and his friends could start all the petitions they wanted, but the people in power weren't obliged to listen. The *New York Times* might call Aaron Swartz a folk hero, but to the federal government, Swartz was still just *a big hacker.*

Swartz's indictment roughly coincided with the twentieth anniversary of the World Wide Web, which had been announced to the world on August 6, 1991. Rather than retain the rights to his invention, Tim Berners-Lee had ceded control to the public, and the public had become the source of its power. Now the Web encompassed everything from academic authors lobbying for open access to amateur pornographers uploading homemade smut to the website YouPorn. It was Project Gutenberg, and Eldritch Press, and a million tart gossip blogs, and a billion uninformed website comments. It was Tunisians and Egyptians using Twitter to coordinate protests during the Arab Spring; it was Tunisians and Egyptians using Facebook to waste time at work.

Decentralized by design, the Web was a tool that rapidly animated and disseminated ideas good, bad, and between. That most people seemed only to use it for banter and ephemera was no demerit; the medium's discursiveness was a sign of its strength. The Web had no presiding officer to direct debate and impose decorum. The Web belonged to its users.

Yet, since its beginnings, government and industry consistently acted as if the Web were a hierarchical system to control and subdue, from which deviant perspectives could definitively be excommunicated. COICA and PIPA were products of the same reactionary mentality that had produced the NET Act, the Digital Millennium Copyright Act, the Computer Fraud and Abuse Act, and other such statutes: a mentality that reflexively discouraged ideas, behaviors, and interactions that it neither recognized nor understood. "I, personally, do not think the world at large really, sincerely wants to provide literacy and education from anyone to The Third World, in spite of all lip service to the contrary," a morose Michael Hart wrote to friends in July 2011.[57] He died of a heart attack two months later, at sixty-four, with the electronic renaissance he had predicted decades earlier as near and as distant as ever.

In a December 2010 *Scientific American* article titled "Long Live the Web," Berners-Lee warned that while the Web had "evolved into a powerful, ubiquitous tool because it was built on egalitarian principles," malign forces were actively working to erode those principles. "Large social-networking sites are walling off information posted by their users from the rest of the Web. Wireless Internet providers are being tempted to slow traffic to sites with which they have not made deals. Governments—totalitarian and democratic alike—are monitoring people's online habits, endangering important human rights. Why should you care? Because the Web is yours."[58]

ON October 26, 2011, Representative Lamar Smith of Texas introduced into the House of Representatives the Stop Online Piracy Act, or SOPA, an industrial-strength (and industry-approved) version of PIPA. Although it was framed as another weapon in the war against overseas content pirates, SOPA was so broadly constructed that it could theoretically have been used to inhibit innocuous online activity in the name of copyright protection. "SOPA would undermine all of the best parts of the Internet, forcing sites that relied on user-generated content to police that material before it ever even made it online," David Segal observed. "Foreign sites would have to prevent certain content from being uploaded or risk being blocked from American view. Domestic sites would have to scrub out any links to such blocked sites."[59]

Swartz and his allies found the bill abhorrent. But, like its predecessors, SOPA also seemed likely to pass. By November 2011, SOPA and PIPA had won the support of many groups: the Motion Picture Association of America, the US Chamber of Commerce, several major labor unions, the publisher Reed Elsevier, and many others. No significant public resistance or congressional opposition had emerged. Though certain tech companies, notably Google, had contributed money to anti-SOPA lobbying efforts, they were late to the game; in his book *The Fight over Digital Rights*, Bill D. Herman noted that, through mid-November 2011, proponents of SOPA and PIPA outspent the bill's opponents by a six-to-one margin.[60]

On November 16, 2011, the House Judiciary Committee held hearings on SOPA. Six witnesses were called to testify; the first was Register of Copyrights Maria Pallante. "Congress has updated the Copyright Act many times in the past two hundred years, including the enforcement provisions, but as we all know, this work is never finished. Infringers today are sophisticated and they are bold," Pallante said. "This is not a problem that we can accept."[61]

The other witnesses' testimonies were of much the same tenor, with the exception of Katherine Oyama, from Google, who warned that, if SOPA became law, "countless websites of all kinds, commercial, social, personal, could be shuttered or put out of business, based on allegations that may or may not be valid, and the resulting cloud of legal uncertainty would threaten new investment, entrepreneurship, and innovation."[62]

The committee members peppered Oyama with a line of combative questions indicating their frustration at the perceived extent of the online piracy problem, and at Google's perceived reluctance to stop it. "You get page after page of free *Grinches*," groused the Texas congressman Ted Poe, describing the results of a Seussian Google search gone horribly, piratically wrong.[63] Poe spoke for many in the room when he refused to characterize online copyright infringers as *bad actors*: "They are not bad actors, they are thieves. And this legislation is trying to get a grip on this."[64]

In his Safra Center paper on the workings of Congress, Swartz had written of the process by which representatives become alienated from the interests and opinions of their constituents, and how experts and lobbyists insulate legislators from the prevailing winds. The arrival of SOPA and the continued existence of PIPA seemed evidence enough that dissenting voices still hadn't been heard. Swartz and his cohorts would need to try different tactics.

Earlier that year, Swartz's old friends Tiffiniy Cheng and Holmes Wilson had founded an organization called Fight for the Future, dedicated to raising opposition to SOPA and PIPA. With characteristic élan, they had cooked up an idea they called Internet Censorship Day, which would take place on the date of the House Judiciary Committee hearing. On Internet Censorship Day, webmasters would "censor" their own websites by replacing their logos or home pages with a banner reading "Stop Censorship," or a larger graphic that featured an ominous message: "WEBSITE BLOCKED."

"Pursuant to HR 3261 (SOPA) this website has been blocked to persons in the United States," the latter graphic read. "Sound scary? Today, Congress holds a hearing on a bill that would create America's first system for internet censorship."[65] Both images then linked to a form through which Web users could contact Congress and ask their representatives to "reject the Internet Blacklist Bills."

Swartz built the contact-Congress tool for Fight for the Future and helped spread the word about Internet Censorship Day. The tool launched on November 16, with hundreds of websites—including popular sites such as Reddit, 4chan, and Tumblr—blacking themselves out and encouraging stymied Web users to contact Congress and complain. According to Fight for the Future's own estimates, the form that Swartz built generated approximately 1 million e-mails to Congress on November 16. Legislators' telephones rang constantly that day. Yet, in the end, only a few legislators changed their positions on SOPA and PIPA. "INTERNET, YOU ARE AMAZING," Cheng and Wilson wrote on the Internet Censorship Day website. "But SOPA's still alive. Prepare for Round 2."[66]

Swartz, who had relocated to New York from Cambridge, was

working with the global activist organization Avaaz and assisting Ben Wikler with the development of a new podcast, to be called *The Flaming Sword of Justice*, about progressive politics and political activism. "When he first moved to New York, he hated it," Stinebrickner-Kauffman recalled. She was in New York that November, too, cohabiting with Swartz in a sublet on the Upper West Side to see what domestic life together might be like.

"I think partly he felt he needed to make a clean break because of the indictment. He wanted to get away from Cambridge," she said. But he couldn't escape altogether. The terms of Swartz's bail required him to report in person to the court in Boston once every two weeks, so every other Monday Swartz boarded an early-morning bus and traveled to Massachusetts to demonstrate that he hadn't fled.

As best he could, Swartz continued his work on behalf of the stop-SOPA coalition, trying to entice new allies to the cause. "I remember at one point during this period, I helped organize a meeting of start-ups in New York, trying to encourage everyone to get involved in doing their part," Swartz later related. "And I tried a trick that I heard Bill Clinton used to fund his foundation, the Clinton Global Initiative. I turned to every start-up founder in the room in turn and said, 'What are you going to do?'—and they all wanted to one-up each other."[67]

Cheng and Wilson decided to follow up on the relative success of Internet Censorship Day with another, bigger Internet blackout, which they scheduled for January 18, 2012, six days before the Senate was scheduled to vote on SOPA and PIPA. As before, participating websites would replace their logos or home pages with a stop-SOPA message and a tool that would help visitors contact Congress.

This time around, many of America's most popular websites joined in. Longtime champions of open culture such as Wikipedia, WordPress, Craigslist, and Reddit turned their sites entirely black. Some commercial giants—Pinterest, Google, Amazon, eBay—shut down as well, along with approximately 115,000 others by Fight for the Future's count, their participation serving to illustrate the "darkening effect" that SOPA and PIPA would have on the Internet at large. "The Internet Archive is already blacklisted in China—let's prevent the United States from establishing its own blacklist system," wrote Brewster Kahle.[68] "If you like what we do, oppose SOPA and PIPA," Project Gutenberg said on its Facebook page.

The unexpected inaccessibility of some of their favorite websites caused lots of people—including many members of Congress—to take notice of SOPA and PIPA for perhaps the first time. Millions of people reached out to Congress to register their opposition to the bills, via e-mail and over the telephone, and legislators scrambled to assure their constituents that their concerns had been registered. "#NJ: I hear your concerns re: #PIPA loud & clear & share in these concerns. I'm working to ensure critical changes are made to the bill," New Jersey senator Bob Menendez posted on Twitter.[69] "Thanks for all the calls, emails, and tweets. I will be opposing #SOPA and #PIPA. We can't endanger an open internet," tweeted Oregon senator Jeff Merkley.[70] "Freedom of speech is an inalienable right granted to each and every American, and the Internet has become the primary tool with which we utilize this right," Illinois senator Mark Kirk said in a press release. "This extreme measure stifles First Amendment rights and Internet innovation. I stand with those who stand for freedom and oppose PROTECT IP, S.968, in its current form."[71]

The bills were tabled two days later, on January 20. "I welcome [this] announcement because of the bill's overreach," wrote Representative Ted Poe, whose heart grew three sizes that day.[72] (Not every former supporter had been flipped. In a press release, Senator Patrick Leahy warned, "The day will come when the Senators who forced this move will look back and realize they made a knee-jerk reaction to a monumental problem.")[73] It was just about nine years to the day since the Supreme Court had ruled against Eric Eldred in *Eldred v. Ashcroft.* "If we cannot overturn it in the courts, then we shall overturn it in the legislatures," the sixteen-year-old Swartz had

predicted at the time. The Copyright Term Extension Act still survived as the law of the land. But SOPA and PIPA had been strangled
in the cradle. For once, the legislatures had stood down.

A few days later, Swartz appeared on Ben Wikler's podcast, *The
Flaming Sword of Justice*, to discuss the significance of the SOPA and
PIPA victories. "It was so incredibly inspiring," Swartz told Wikler.
"People often think, like, 'Oh, I sign these petitions,' you know, 'I
don't really know what effect they're having, does anyone listen to
them?' This makes it really clear. People *are* listening, and when we
all speak up, we can totally change the debate. We shifted the entire
landscape of this issue."[74]

Even if this sentiment was overstated, Swartz and his allies deserve significant credit for their work. Not only did they organize
the blackouts and protests and galvanize public opposition to the
bills, they also encouraged ordinary Internet users to think about the
political implications of the medium. As Holmes Wilson put it later,
"Demand Progress was the first organization to build campaigns
that connected those bills to entire new audiences of people who
cared about tech policy just because of how much they lived on the
Internet, and not because of any previous kind of commitment to
the ideals of liberty online."[75]

Swartz made a similar point during his interview with Wikler,
in an attempt to articulate the broader meaning of the SOPA and
PIPA protest movement: "You know, [Motion Picture Association
of America chairman] Chris Dodd gave a nice interview to the *New
York Times* where he kind of explained his whole evil master plan
for sneaking this bill through before anybody noticing, and he said,
he was so upset, you know, it was, like, 'Those meddling kids, they
stopped us! This isn't supposed to happen!' He said, 'In my 40 [*sic*]
years in the Senate I've never seen anything like this. This is, like, the
Arab Spring has come to the Internet.' And I think that's what it is.
These people are running scared, and we need to continue the fight."[76]

That spring, Stinebrickner-Kauffman and Swartz moved in to-

gether for good. Swartz used the time he had previously been spending on SOPA and PIPA to explore a variety of new interests. In
conjunction with the photographer Taryn Simon, he developed an
art project called Image Atlas, which gathered and juxtaposed the
top image-search results for common search terms—such as *love*
or *freedom*—from fifty-seven countries around the world; the piece
was eventually exhibited on the website of the New Museum. He
had started unraveling David Foster Wallace's notoriously dense
novel *Infinite Jest*. "He spent, like, entire weekends where he was
mostly working on this plot summary of *Infinite Jest*," Stinebrickner-
Kauffman recalled. "He loved taking complex narratives and distill-
ing their essences."[77]

In May 2012, Swartz traveled to Washington, DC, to speak at
the annual Freedom to Connect conference. There, he distilled the
complex narrative of the stop-SOPA movement into a compelling
drama. Swartz warned the crowd of democracy activists that it was
too early to celebrate, that similar legislation would recrudesce in
the future:

> Sure, it will have yet another name, and maybe a different excuse,
> and probably do its damage in a different way. But make no mis-
> take: The enemies of the freedom to connect have not disappeared.
> The fire in those politicians' eyes hasn't been put out. There are a
> lot of people, a lot of powerful people, who want to clamp down
> on the Internet. And to be honest, there aren't a whole lot who
> have a vested interest in protecting it from all of that.[78]

Back when he blogged about Kafka's *The Trial*, Swartz cited its
lesson that there's no beating a bureaucracy through official channels, that unexpected stratagems are the only way to get what you
want in such a setting. Swartz observed that K., Kafka's protagonist,
"takes the lesson to heart and decides to stop fighting the system and
just live his life without asking for permission." Swartz had come to

the same conclusion and had lived that way for a while, too. Engaging with bureaucracies on their own terms gets you nowhere—the best course is to disregard their rules and follow a different path.

But left unmentioned in Swartz's post was how *The Trial* ends. K. is visited by two pale, silent gentlemen, clad in black, who join his arms and march him out of his house, through the town, and to a desolate quarry. "Was he alone?" K. wonders. "Was it everyone? Would anyone help? Were there objections that had been forgotten?" Then the two government agents unsheathe a butcher's knife, grab K. by the throat, and stab him through the heart.

10

HOW TO SAVE THE WORLD

On July 28, 2011, a week and a half after being indicted, Aaron Swartz posted to his website a working draft of a document modestly titled "How to Save the World, Part 1."[1] The post, which he shared with only a few friends, synthesized many of his insights into the ways in which the Internet could be used to organize and catalyze meaningful social change. It was an optimistic theme, all things considered. At the time, for all the talk about how online activism could help save the world, Swartz had little firsthand evidence that such an outcome was possible. He had posted the "Guerilla Open Access Manifesto" online, and the orthodox open access community had recoiled from it. Despite Swartz's best efforts to rally opposition to PIPA, the bill had passed unanimously in the Senate Judiciary Committee and, as of July 2011, looked as if it would soon become law. He had worked to help elect progressive Democrats during the 2010 midterm elections; the Democrats ended up losing control of the US House of Representatives.

But the revolution will be A/B tested, and maybe even A/B/C/D and E tested, too. In his working paper, Swartz described his new plan for the future of activism. Rather than form a political action group focused on one single issue or tactic, Swartz proposed that organizers should assemble groups of people supremely competent in certain relevant disciplines—investigators, activists, lawyers, lobbyists, policy experts, political strategists, journalists, and publicists—who could combine their efforts and advocate effectively for *any* issue, big or small. Swartz envisioned a flexible, intelligent, multifaceted task force that would learn from its mistakes and refine its tactics accordingly: a team of specialists that, cumulatively, worked as generalists.

The informal coalition that assembled to successfully protest SOPA and PIPA roughly conformed to Swartz's vision of a polymathic activist committee. Their success, in a sense, served as a proof of concept. In the spring and summer of 2012, as the federal case against him continued, Swartz spent his time refining that concept further and mulling the best way to build an organization that did not disempower its members.

If Swartz's model organization already existed, he hadn't found it yet. He certainly hadn't been impressed by most of the large systems he had encountered over the years. He had dropped out of high school and college, forced his own departure from Reddit, and left a string of half-finished projects and short-lived collaborations behind him. Since relocating to New York, he had worked briefly at Avaaz, even more briefly at the activist group Change.org, and most recently at the global IT consultancy ThoughtWorks, a job that he seemed to enjoy—but who could tell, really? Swartz's restlessness had earned him a reputation. "Because he moved from thing to thing fairly rapidly, that actually ended up imposing a cost, so that people were wary about starting a really big project with him because they worried he wouldn't stick around," said Ben Wikler. "That was a reasonable concern to have."[2]

But just as the advice of an escape artist can be useful in building better jails, Swartz's estrangement from large organizations provided him with meaningful insight into their defects. He wasn't, however, quite sure yet about the most efficacious way to fix them—not in a sustainable manner, at least. He searched everywhere for solutions. "He was reading all these management books," said Wikler. "He was reading biographies of Sam Walton and all these people who built big organizations. He spent hours talking about management and how to make teams work."[3]

Swartz sought out examples of particularly benevolent and well-managed institutions. "What organizations have the mission of doing the most good possible? So far I've found: @Avaaz, @Change, and @GiveWell. Any others?" he posted to Twitter on June 1, 2012.[4] That same month, on his blog, Swartz bestowed effusive praise on Hollister Co., marveling at the chain clothier's success at creating a distinct retail experience that was replicated seamlessly from store to store. Every Hollister outlet featured "the perfect music" and "the perfect scent" and "a perfect photograph for the wall." Somehow, Hollister had built "five hundred stores of perfection," he wrote. "(You may detest what they are perfect at, but that's not my point. The point is that they have a vision and make it stick.) How do they do it?"[5] Selling cheap clothes, saving the world: both required a vision, an organization, and the ability to imbue the latter with the former.

During the spring and summer of 2012, while Swartz was dreaming of the perfectly efficient organization, he remained mired in judicial and academic bureaucracies. His legal situation evoked one of M. C. Escher's recursive drawings, characterized as it was by infinite repetition and the illusion of forward momentum. "I would ask him periodically what's going on with the case," Taren Stinebrickner-Kauffman recalled.[6] "And for a very long time the answer was 'Nothing, because the government won't turn over the documents that they're supposed to turn over.'" MIT had been no more responsive than the prosecutors. Between May and August 2012, Swartz's attorneys contacted the institute's outside counsel thirteen separate times to request a meeting with MIT officials, in hopes of convincing them to issue a public statement that could be used on Swartz's behalf. MIT officials did not respond.[7]

With official communications blocked, Swartz's father decided once again to convene a meeting with MIT executives through informal channels. With the help of Media Lab director Joi Ito, Robert Swartz sent a letter to MIT's president and chancellor on August 10, 2012. "We think there are both legal and non-legal issues that you are not aware of and urgently ask for a meeting," he wrote. "The urgency of the meeting is due to the fact that the prosecutor has given us a deadline of Wednesday [August 15] to resolve the case or go to trial and we have a meeting Monday with the head of the criminal division that requires hard decisions."[8]

The letter was an epistolary kowtow. But even though MIT's president read the letter soon after it was sent, the school did not meet with Robert Swartz for another month. Aaron Swartz would have to face the August 15 deadline on his own.

As the date approached, Swartz evinced public calm, posting on Twitter about the US presidential campaign and his favorite podcast-listening tools, and blogging about the flaws of "scientism": the bias that only scientific methods have validity. "If you're struggling with a decision, we're taught to approach it more 'scientifically,' by systematically enumerating pros and cons and trying to weight and balance them," Swartz wrote on August 10. "Well, studies have shown that this sort of explicit approach repeatable [sic] leads to worse decisions than just going with your gut."[9]

In private, though, Swartz was thinking hard about whether to resolve his case via plea bargain or proceed to trial. He kept his own counsel. "I didn't know the details of the deal, I didn't have the context, and it was really hard because I didn't know what was going on with him," Stinebrickner-Kauffman remembered. "I knew that there was a plea deal, but he just sort of shut me out."[10]

Swartz ultimately rejected the deal, and, later, he and Stinebrickner-

Kauffman quarreled over his willful mutism. "I was, like, 'You can't shut me out of this because it's too important for me,'" Stinebrickner-Kauffman recalled. She was right—and Swartz knew it. As he matured, Swartz began to realize that "Aaron stands alone" was a suboptimal life strategy, one that bred isolation and pain both for himself and for his loved ones. He'd come to see that he didn't have to operate like the ineffective systems he so despised, in which bad habits calcified and soon became the norm. He could choose to refine his tactics and learn from his mistakes. He could choose to elevate, even as his surrounding circumstances kept trying to drag him down. His own life was the one system he *could* control.

Three days after the August 15 deadline expired, Swartz launched a blog series titled Raw Nerve, as in "you've struck a." The series covered a relevant topic: daily life, and how to get better at it. "After all, in my day job, I'm constantly looking for ways to learn and grow—reading the latest books and articles about the field, talking to other people with similar jobs and hearing what's worked for them," he announced in the first entry. "Why aren't I doing the same thing for life?"[11]

While Swartz had always experimented with life-hacking strategies—"I had some time to kill this morning so I decided to work on improving my eyesight," he had blogged as an aspiring fifteen-year-old homeopath[12]—Raw Nerve was his most ambitious attempt yet at articulating and systematizing this type of advice. Full of pithy life lessons drawn from sociological research papers, the Raw Nerve posts evoke the writings of Malcolm Gladwell; Swartz even based one post on the same study that Gladwell used as the basis for his 2008 book *Outliers*. But Raw Nerve also transcended pop-science superficiality, and for one basic reason: Swartz clearly intended its motivational homilies, first and foremost, for his own benefit.

One of the self-improvement strategies that Swartz articulated was to *believe you can change*: deny your own immutability, refuse to accept that the future is dictated by the past.[13] Swartz had always been awkward, secretive, and self-reliant, but there was no reason to treat those traits as if they were a terminal condition. He gradually opened up to Stinebrickner-Kauffman and stopped excluding her from his deliberations.

Another strategy was to *take a step back*: choose to view life from a broader perspective rather than getting stuck in daily minutiae. Acknowledging the bigger picture could be an empowering choice. "I feel more in control of my life, more able to cope with my problems. I feel like I'm charting my own destiny, instead of following some track," he wrote. "I feel like I'm growing as a person."[14]

Perhaps the most difficult lesson of Raw Nerve was that life requires us to *lean into the pain*: refuse to shirk difficult, uncomfortable tasks and situations.[15] In late summer 2012, Swartz confronted one of his biggest bogeymen: asking other people for help. He had been fighting the charges with his own money, and it was running out. His friend Bettina Neuefeind—an attorney who was married to Lawrence Lessig—set up a legal defense fund. Swartz himself started soliciting contributions from wealthy friends and acquaintances. That August, Jeffrey Mayersohn, owner of the independent Harvard Book Store in Cambridge, Massachusetts, received an e-mail from Swartz asking for a donation. Mayersohn agreed to donate to the fund and offered to organize more fund-raising events. Swartz accepted Mayersohn's contribution, but declined the offer of further help. "It was hard enough for me to make this simple ask of you," he replied, according to Mayersohn.[16]

Asking outsiders for help was still a torturous task, but Swartz had faced his fear, rather than run from it, and that was a significant development for a man with such a strong flight reflex. No one could change all of their bad habits at once, but incremental changes accumulate, and forced habits can be unlearned. "Next time you start feeling that feeling, that sense of pain from deep in your head that tells you to avoid a subject—ignore it," Swartz wrote.[17] The world might never stop trying to hurt you, but *you* were allowed to set your own threshold for pain.

ON September 12, 2012, the US Attorney's Office in Boston filed a new indictment against Aaron Swartz. Federal criminal indictments cannot be amended once they are filed, so if a prosecutor wants to add new charges, he must file a superseding indictment, which replaces the old one. New information or new developments in a case commonly account for superseding indictments. But the only thing that had changed since the first indictment came down was that Swartz had shown himself willing to vigorously contest the charges.

This superseding indictment, like many sequels, doubled down on the elements that had made its predecessor such a hit. It increased the number of felony charges against Swartz from four to thirteen, and the maximum number of years he faced in prison from thirty-five to ninety-five. As an acquaintance of Swartz's, Seth Finkelstein, put it on his blog: "as I've said before—they don't like him. They really don't like him."[18] If that point hadn't been made clear before, the superseding indictment left little doubt.

In the first indictment Swartz had been charged with one count each of wire fraud (maximum prison sentence: twenty years), computer fraud (maximum prison sentence: five years), theft of information from a computer (maximum prison sentence: five years), and recklessly damaging a computer (maximum prison sentence: five years). The superseding indictment now charged him with one count of recklessly damaging a computer (maximum prison sentence: five years), two counts of wire fraud (maximum prison sentence: forty years), five counts of computer fraud (maximum prison sentence: twenty-five years), and five counts of unlawfully obtaining information from a protected computer (maximum prison sentence: twenty-five years). All but the wire-fraud charges were brought under the Computer Fraud and Abuse Act. The government manufactured these new counts by treating the downloading incidents in September, October, November, December, and January separately—a seemingly gratuitous prosecutorial tactic that reeked of menace and intimidation: "You want to have a trial? Fine, let's have a trial."

The Department of Justice counsels US attorneys against filing unnecessary charges against defendants, in the interests of expediency and justice. "The bringing of unnecessary charges not only complicates and prolongs trials, it constitutes an excessive—and potentially unfair—exercise of power," notes the *United States Attorneys' Manual*, which advises federal prosecutors on best practices.[19] The US Attorney's Office in Boston held no press conference to explain its decision to bring these new charges, but on his website the day after the superseding indictment was announced, Swartz, quoting the philosopher Bertrand Russell, offered an oblique theory on the prosecutors' motives: "Since power over human beings is shown in making them do what they would rather not do, the man who is actuated by love of power is more apt to inflict pain than to permit pleasure."[20]

The same day that the superseding indictment was filed, Swartz's father finally met with MIT's chancellor and general counsel. But he left that meeting feeling as stymied as ever. "First, [Robert Swartz] wanted MIT to make a public statement. This, the Chancellor and General Counsel explained, MIT would not do," a report noted of the meeting.[21] *Confront reality* was one of the life strategies Swartz had articulated in Raw Nerve. The reality of his legal situation at the time was that the tactics he had deployed thus far had been ineffective. The government was still intent on jailing him; MIT was still intent on ignoring him. When Swartz had first been indicted, he'd responded by starting an online petition. This time—A/B testing, etc.—Swartz hired a new lawyer.

At the end of October 2012, Swartz discharged Martin Weinberg and retained Elliot Peters, a partner at the boutique San Francisco firm Keker & Van Nest, as his lead counsel. (Weinberg had taken over the case from Swartz's first attorney, Andrew Good, in November 2011.) As the case consumed more and more of his time, Swartz had ceased posting to his blog. After concluding the Raw Nerve series on September 25, he wrote only two more entries all year: an essay about the movie *Looper* on October 8, and another on the film *The Dark Knight* on November 1.[22] *The Dark Knight* is a superhero film in which Batman, the villainous Joker, and district attorney Harvey Dent vie for the soul of Gotham City. Each character offers a different strategy for saving the city from organized crime and political corruption. Batman embraces masked vigilantism. Harvey Dent has scores of gangsters arrested on spurious charges. "The Joker had by far the most interesting plan," wrote Swartz: "he hoped to out-corrupt the corrupters, to take their place and give the city 'a better class of criminal.'"

In his essay, Swartz strongly supported the Joker's policy platform. Although the Joker presents himself to the world as a deranged and murderous clown, Swartz claimed that the Joker is actually *"homo economicus,"* a supremely rational actor, the character who best understands both the problems facing Gotham City and the best solutions to those problems. Batman might have had better gear and Harvey Dent might have had the public's sympathy—but the Joker understood game theory, the best weapon of all.

Though the Joker's methods—such as burning large piles of money and blowing up hospitals—might have been controversial, the logic behind them was sound. "And the crazy thing is that it works!" Swartz enthused. Not only did the Joker end up ridding the city of organized crime, he convinced Gotham's residents to reevaluate their world and their roles in it. "The movie concludes by emphasizing that Batman must become the villain," Swartz wrote, "but as usual it never stops to notice that the Joker is actually the hero."[23]

ELLIOT Peters's counsel led to a shift in Swartz's legal strategy. Swartz's previous attorney, Martin Weinberg, had focused on the potential privacy violations in the packet capture and sought to suppress evidence the government had gathered by means of allegedly improper searches. Peters continued to argue for these motions, but prosecutors' responses revealed their disdain for this line of argument. "Apparently without a trace of irony," they wrote in November 2012, "Swartz argues that MIT and law enforcement violated his rights to privacy as he hid his computers and hard drives in MIT's locked wiring closet, used pseudonyms to avoid identification, hardwired his computers to MIT's network switch to avoid detection, siphoned off JSTOR's copyrighted documents, kept reconfiguring his computer to circumvent MIT's and JSTOR's efforts to keep him off their networks, and relocated the evidence to MIT's student center."

So Peters opened up a new front for Swartz's defense. He began to argue that the "unauthorized access" charges were misapplied. Swartz hired an expert witness, a computer security consultant named Alex Stamos, who contended that MIT's computer network was so open that Swartz's access in no way could have been construed as unauthorized. Stamos later noted in a blog post that "in my 12 years of professional security work I have never seen a network this open," and that JSTOR had "lacked even the most basic controls to prevent what they might consider abusive behavior."[24] Stamos claimed that it would have been easy for MIT to limit guest access to JSTOR, and that its failure to implement such security measures even after Swartz's downloads were detected could only be construed as tacit authorization of his presence.[25]

As the weeks went by, Peters grew more optimistic about the prospect of suppressing evidence taken from the laptop and USB drive that police had seized when Swartz had been arrested. On December 14, Peters and Heymann met with Judge Nathaniel M. Gorton to review the status of the suppression motions filed by Swartz's defense team. Judge Gorton granted Peters's request for an evidence-suppression hearing and scheduled it for January 25, 2013. "If we had won that motion and suppressed the fruits of their search, they wouldn't have had a lot of the evidence they had planned to use at trial," Peters remembered.[26] Afterward, still in the courthouse, Taren Stinebrickner-Kauffman moved to embrace Swartz, but was rebuffed. "Not in front of Steve Heymann," Swartz told her. "I don't want to show Steve Heymann that."[27]

Perhaps feeding off Peters's energy, Swartz took steps to expand his circle of supporters and get more people invested in his case. Jeffrey Mayersohn got another e-mail from Swartz that December, asking if he'd be willing to meet with Charlie Furman, whom Swartz had hired to help coordinate fund-raising efforts for his legal expenses. "And when that happened, I was really encouraged because I thought, 'Okay, this is Aaron getting engaged in his defense,' " Mayersohn said.[28]

But leaning into the pain didn't make it stop hurting. Swartz hated having to ask people for money because doing so meant having to discuss the details of his case. "The more that people talked to him about the case instead of talking to him about their exciting technology ideas, or their ideas about how to change the world, the harder that was for him in a lot of ways, I think," Stinebrickner-Kauffman suggested.[29]

The case cut into the time he could spend on his many projects as well. Though his productivity had been sapped, he still tried to keep as busy as possible. As 2012 drew to a close, he was preparing to revive watchdog.net. He was conducting substantial research on how to reform drug policy. At ThoughtWorks, he was leading a team that was developing new digital tools for political activism and had embraced the challenges of personnel management. At a company retreat in Chicago, he gave a presentation about the Toyota lean-production system. When he returned to Brooklyn, he was so excited about the experience that he gave an encore performance for Stinebrickner-Kauffman.

Swartz may have been reluctant to show emotion in front of Heymann, but he was growing more and more comfortable with showing it at home. November and December 2012 marked the zenith of their relationship, Stinebrickner-Kauffman recalled. "Being with you just keeps getting better," Swartz told her at one point.[30]

Swartz and Stinebrickner-Kauffman rang in 2013 with a New Year's vacation in Burlington, Vermont, accompanied by Quinn Norton's young daughter, Ada. It was the first time in a while that Swartz had taken a break from his case. Unfortunately, he contracted the flu, which kept him indoors for most of the trip, and the virus persisted when he returned to New York. "He asked me a couple of times that week, in a voice that could have been rhetorical or flippant, and I thought was rhetorical or flippant at the time, 'Am I always going to feel like this?'" Stinebrickner-Kauffman remembered. "And I thought he was talking about the flu."[31]

He recovered sufficiently to travel with Stinebrickner-Kauffman to Holmes, New York, on Wednesday, January 9, for a meeting of multi-issue political activists from around the world. Swartz's friend Sam McLean, a political organizer from Australia, also attended the conference, and the two men separated from the larger group and secluded themselves in conversation, discussing ways to build intelligent systems that would empower ordinary citizens to challenge entrenched corporate and partisan interests. "His argument was that we'll have to fight more SOPA-style campaigns. So we need an algorithm or computer program that would encourage lots of people to identify the fights and to start the campaigns," McLean told the *Sydney Morning Herald* in 2014. "We'd put the tools that we have at our disposal in their hands."[32]

Swartz had actually been building tools like these for several months with his colleagues at ThoughtWorks. Victory Kit, as the project was called, was an open-source version of the expensive community-organizing software used by groups such as MoveOn. Victory Kit incorporated Bayesian statistics—an analytical method that gets smarter as it goes along by consistently incorporating new information into its estimates—to improve activists' ability to reach and organize their bases. "In the end, a lot of what the software was about was doing quite sophisticated A/B testing of messages for advocacy," remembered Swartz's friend Nathan Woodhull.[33]

Swartz was scheduled to present Victory Kit to the group at the Holmes retreat. At the last minute, though, he declined to do so. "I've thought at that gathering that he acted a little oddly, in retrospect, but at the time I thought it was just Aaron being Aaron," Woodhull said. Swartz's talk apparently depended on finding a partner to join him in making the code open-source, and when Swartz couldn't secure a commitment in time, he decided not to deliver his presentation.

On the drive back to the city, Swartz held forth on the inefficacy of most nonprofit organizations; they were designed to deploy their tactics without bothering to evaluate whether those tactics actually worked, he grumbled. Woodhull later noted that Swartz was fascinated by "this whole idea that there are all these organizations that think they are doing good, but actually, in objective terms, either aren't or could be doing a lot better." Too many nonprofits took the Batman and Harvey Dent routes, so to speak: they deployed methods that were emotionally satisfying but empirically ineffective. What the nonprofits needed to do was emulate the Joker. If you removed emotion and presumption from your strategies and approached your problems rationally and analytically—if you learned from your mistakes and refined your tactics accordingly—then you had a much better chance of devising the optimal solution to any given problem, no matter how crazy that solution might seem.

That same afternoon, Peters called Heymann to discuss the upcoming evidence-suppression hearing. "Toward the end of it," Peters recalled, "I said, 'Can't we find some way to make this case go away?' I remember saying to them, 'It's just not right for this case to ruin Aaron's life.'"[34] Heymann responded with a familiar refrain: the government would never agree to a deal that didn't include jail time, and if Swartz was convicted at trial, the prosecution would seek a guidelines sentence of at least seven years in prison. You want to have a trial? Fine, let's have a trial.

Though Swartz returned to New York on Wednesday, Stinebrickner-Kauffman remained at the retreat overnight. She returned to Brooklyn on Thursday evening and was surprised to find Swartz at home, on the couch, in an unexpectedly social mood. "Surprise!" he exclaimed, and despite Stinebrickner-Kauffman's fatigue, he insisted they go out and meet some friends in Manhattan's Lower East Side neighborhood, at a bar called Spitzer's Corner. They were hungry, so upon arrival they ordered two of Swartz's favorite dishes, grilled cheese and macaroni and cheese. They agreed the grilled cheese was one of the best they had ever had.

When Swartz woke up the next morning, he was sluggish and gloomy. "I couldn't get him out of bed. He was just in a really terrible mood. Just despondent about the case and everything," Stinebrickner-Kauffman recalled. As she got ready for work, she tried various tactics to lift his spirits—opening windows, playing music, tickling him. Eventually, he rose and got dressed, donning a black T-shirt and corduroy pants. "I got ready to go, I was running late for a meeting, and he said he wasn't going to come with me. And he was going to stay home and rest," Stinebrickner-Kauffman remembered. "And I asked him why he had gotten dressed. And he didn't answer."[35]

When she got to work, she found Ben Wikler online and told him that she was worried about Aaron. Wikler initially suggested that Swartz work from Wikler's apartment that afternoon. He and his wife had recently had a baby, a boy, and Wikler thought that working in proximity to the newborn might lift Swartz's spirits. Instead, they settled on sharing dinner together that night. They tried to call Swartz, but his phone was off—"which was not super-unusual for him," Wikler said.[36] The day went by, amid hopes that all would be right by evening.

That afternoon, in his office in California, Elliot Peters reviewed some of the evidence that the prosecution had handed over in late December.[37] As he read, he got more and more excited about Swartz's chances at the upcoming suppression hearing. "I ran down the hall, saying, 'Look at this! Look at all this!' " he remembered.[38]

Peters put the new evidence in his briefcase, left his office, and hopped into his car. As he drove, his phone rang. It was Robert Swartz. Aaron had killed himself.

STINEBRICKNER-KAUFFMAN found Swartz around 7:00 p.m. on Friday, January 11, hanging from a window in their apartment, his body cold, his belt looped around his neck, still wearing the clothes he had donned that morning. She screamed so loud that emergency response operators could not initially understand what she was saying. Stinebrickner-Kauffman called Wikler almost immediately with the sudden, terrible news. He received her call at 7:05. "That's when everything shattered," said Wikler.[39] Over the next few days, Swartz's friends and relatives were left to reassemble the various pieces.

Swartz's death became a national story faster than anyone could have expected, especially considering the relatively meager coverage his ongoing case had received while he was alive. The mourning soon spread worldwide. On a memorial website titled Remember Aaron Swartz, which had been hastily put together by some of his friends, mourners from around the globe posted messages of remembrance and sympathy. "I met Aaron twice and, each time, I was struck by the searing lucidity of his mind, by his uncanny ability to see further than most of us," wrote Jean-Claude Guédon, a Canadian academic who had attended the EIFL conference in Cupramontana with Swartz in 2008.[40] Brad Templeton of the Electronic Frontier Foundation reflected that he knew many people who were willing to stand up for their core beliefs, but "most of them have been very strong, more outgoing personalities, who could react to attacks without flinching.

It is perhaps more brave to take on the world when you are shy and it can hurt you so badly."[41] The condolences from strangers outnumbered those from his friends and associates. "I never met him. I am sad," wrote Reynaldo Guimaraes of Brazil. "God bless him and his family."[42]

On the clear, mild morning of January 15, 2013, hundreds of mourners filed past Aaron Swartz's coffin in the parking lot of the Central Avenue Synagogue in Highland Park, Illinois. Television news vans waited on the street outside. Friends, family members, and admirers filled the synagogue to capacity. Some had to stand after all the seats filled up. They had come, representatives from all stages of Swartz's life, to grieve and console and draw strength from one another's presence, as if, as a group, they could make sense of this bad thing.

Inside, some of the people who had known Swartz best addressed the crowd: Lawrence Lessig, Ben Wikler, Taren Stinebrickner-Kauffman. Their remarks shared an angry, confused sorrow over the unnecessarily abrupt end to Swartz's short life. Swartz had always loved explaining things to people, so it was odd that he left no note, no explanation, nothing to clarify why he'd killed himself. Some observers later seized on Swartz's mood swings and suggested that his suicide was the product of undiagnosed mental illness. In his eulogy at his son's funeral, an unshaven, haggard Robert Swartz stifled tears as he offered an alternative theory. "Aaron did not commit suicide," he said, his voice hardening. "He was killed by the government, and MIT betrayed all of its basic principles."

In the weeks and months following Swartz's suicide, this claim became the dominant narrative: the government hounded Swartz to his grave, and MIT sat complacently and watched it happen. On January 13, 2013, hackers attacked the website of the Massachusetts Institute of Technology in retaliation for the school's perceived role in Swartz's death. For weeks, school officials received many angry, accusatory e-mails from Swartz's admirers. "It does not matter how good your reputation is when you find yourself with the blood of a genius on your hands," one correspondent declared. "Not all great Neptune's ocean will suffice to wash it off."[43]

The US Attorney's Office in Boston received its share of public acrimony, too. "As Massachusetts natives, we'll work to end the political careers of the prosecutors here who targeted Aaron," Holmes Wilson promised in a blog post.[44] A person or persons allegedly affiliated with the computer hacking collective Anonymous posted Stephen Heymann's home address and telephone number online, alongside the names of some of his family and friends. "How does it feel to become an enemy of the state?" an unnamed e-mailer asked Heymann soon after this data was released. "FYI, you might want to move out of the country and change your name. . . ."[45] At his home address, Heymann received an ominous postcard featuring a picture of his head inside a guillotine. Elliot Peters told the *Boston Globe* that the day after Swartz killed himself, Stephen Heymann had called and left a message expressing his sympathy. "I can't call him back," Peters told the *Globe*'s Kevin Cullen. "Either I'll say something I shouldn't say, or I'm going to act like I accept his condolences, which I don't. So the only thing I can do is not call him back."[46]

At a memorial service in New York City the weekend after Swartz's funeral, the graphic designer Edward Tufte took the stage and recounted how, as a college student in the 1960s, he had gotten caught hacking the telephone system. Tufte and a friend had built a "blue box"—a device that allowed them to make long-distance phone calls for free—and had used it to make what they believed was the longest-distance phone call ever made: a call from Palo Alto to New York, routed through Hawaii. Six months later A. J. Dodge, a security agent from AT&T, called Tufte and informed him that the company had been onto them the entire time. While the blue-box stunt had technically qualified as a crime, both Dodge and the company's executives realized that Tufte wasn't a criminal, and that nothing was to be gained from treating him as such.

Tufte and AT&T reached an informal resolution, in which Tufte and his friend agreed "that we wouldn't try to sell this . . . we wouldn't do any more of it, and that we would turn our equipment over to AT and T. And so they got a complete vacuum tube oscillator kit for making long-distance phone calls," Tufte said. "But I was grateful for A. J. Dodge and, I must say, even AT and T, that they decided not to wreck my life."[47]

AT&T *decided* not to do it. To pursue a case is always a decision, not an ineluctable gravitational reaction. The government could have given Swartz a break. MIT could have made a statement on his behalf. Swartz could have gone on to greater successes in the future. "He could have done so much more," Robert Swartz said in his eulogy, his voice heavy with regret. "But now he is dead."

Aaron Swartz is dead. And no one has been held accountable. In a March 2013 session of the Senate Judiciary Committee, Texas Republican John Cornyn questioned Attorney General Eric Holder about the prosecutors' conduct and the wisdom of the charges against Swartz, suggesting that it was inappropriate "to try to bully someone into pleading guilty to something that strikes me as rather minor." Holder expressed support for Stephen Heymann, Carmen Ortiz, and the US Attorney's Office in Boston. As of this writing, none of the prosecutors in the Swartz case have been officially censured for their actions. That shouldn't come as a surprise. US Attorney's Offices exist to win convictions, to prosecute alleged lawbreakers without regard to the logic or the justice of the laws that have been broken. Despite Stephen Heymann's resentment of some of Swartz's tactics, the case, in the end, wasn't personal. He was just doing his job, following the rules and customs of the system that employed him.

MIT appointed an independent committee to examine its actions in the Swartz affair soon after his death, amid mounting public anger over the school's inertia. Midway through 2013, the committee released a largely exculpatory report, concluding that MIT had not been unduly negligent or callous in its handling of the Swartz case. It noted that the school was responsible for "respecting its contractual obligations with licensors" and "maintaining the integrity of its network." The report also observed that MIT seemed wholly unprepared for the forcefulness of the public reaction to Swartz's death, especially given the relatively sparse attention the case had received while Swartz had still been alive: "Aaron Swartz's suicide has embroiled MIT in an Internet uproar that the Institute did not anticipate and with which it is not well prepared to grapple as a legal, policy, or social phenomenon."[48]

Swartz's downfall is not so much a tale of personal vendettas and conspiracies as it is a story of flawed organizations. Imperfect institutions cannot help but generate imperfect outcomes. Mistakes compound because systems are designed to produce them. Swartz's seventh and final post in the Raw Nerve series had returned to this theme and asserted that, often, what looked like human error was really systemic failure. The story concerned the Toyota lean-production method that Swartz so admired. In 1982, General Motors closed a plant in Fremont, California, that had been notorious for underproduction and labor grievances. One year later, Toyota came to town, reopened the factory, and rehired the same workers—and, Swartz writes, "so began the most fascinating experiment in management history."

The new Toyota employees were flown to Japan, where they toured Toyota facilities and were introduced to the company's unfamiliar management philosophy. "Three months after they got back to the US and reopened the plant, everything had changed," Swartz wrote. "The Fremont factory, once one of the worst in the US, had skyrocketed to become the best. The cars they made got near-perfect quality ratings. And the cost to make them had plummeted. It wasn't the workers who were the problem; it was the system."[49]

A better system was the solution. "When there's a problem, you shouldn't get angry with the gears—you should fix the machine," Swartz reflected. Unfortunately, some machines resist fixing.

IN December 2001, when he was fifteen, Aaron Swartz went to his website and described an unusual dream: "Last night I had a dream of the way I want to live. I'm not sure it would appeal to other people, but I would certainly like it." He had found himself in a "modernly-designed loft" surrounded by his friends from the Internet, all the hackers and programmers and network enthusiasts who had welcomed a precocious teenager into their ranks: "We were working together on a project that we thought would change the world. We were committed to it, and worked well as a team: we helped each other out with what needed to be done, and kept each other's enthusiasm up." Swartz did not explain what, exactly, they were working on, but the nature of the project seemed less important than the attitude with which they pursued it. "We learned all the time, both finding out the skills we needed to know ourselves, and teaching each other how to improve. While we knew there was a lot to be done, we focused on our one project with a single-minded determination, finishing it and solving all the problems that it raised."

When Swartz woke up, he was still in Highland Park, and still a lonely teenager. But the dream stayed with him. "If anyone wants to send me the money to make this dream a reality, let me know," he concluded. "I already know the people and the project. The problem is we're flung across several continents, and this is the kind of relationship that just doesn't happen over the Internet."[50]

Swartz was wrong on that last point. The idealistic and collaborative organizational dynamic that he described was *more* likely to exist in a digital medium than in the real world. Agglomerative collaboration, in which individual talents are utilized in pursuit of a common goal, animates digital utopianism. It is the soul of the hacker ethic, the motivating ethos of the free software movement.

As Swartz grew up, the Internet delivered both people and projects to him. From the Semantic Web message-board communities to the Creative Commons team, from Open Library to PACER to Demand Progress, all of the world-changing initiatives on which Swartz had worked had been digital. The Internet is both a conduit for idealistic fantasies and a tool that can be used to implement them. It is the real world that often creates resistance.

Throughout Swartz's life, most of his failures and dilemmas derived from his inability to port this dynamic over to a real-world setting. The institutions from which he fled, the shortcuts he took: these were all a result of the real world's fundamental dissimilarity to the Internet. High school and college had been far too hierarchical. He left. Silicon Valley served up more of the same. Systems that were supposed to empower only disempowered. And there's the rub. In the real world, most organizations don't keep learning and improving—they set their parameters and stay within them. Real-world systems aren't outcome focused, they're systems focused.

Swartz never stopped chasing his unusual dream, never became inured to its failure to materialize, and never lost faith in the ultimate plausibility of his vision. By the end of his life, he had realized that it would be up to him to make that dream come true. I MUST INVENT MY OWN SYSTEM OR BE ENSLAVED BY OTHER MEN'S, Michael Hart had posted on his wall in Urbana. Swartz spent his life trying to invent his own systems, ones that were modeled on and powered by the Internet.

Swartz was no guileless technophile. To him, the Internet was not inherently miraculous, nor was digital technology inherently benign. After Apple cofounder Steve Jobs died in October 2011, Swartz drafted an essay depicting Apple as "a ruthless, authoritarian organization" that flouted labor standards and Jobs himself as a martinet who insisted on controlling every aspect of the user experience. His megalomania manifested in Apple's portable music players: sterile white rectangles that could be neither opened nor modified by the end user. "Jobs couldn't abide people opening things," Swartz wrote. "'That would just allow people to screw things up,' he insisted." Swartz saw things differently, and, indeed, devoted much of his life to the notion that the only way that the world ever improved was by allowing people to open things up. This notion—which he shared with Richard Stallman, Michael Hart, the young Noah Webster, and countless other idealists who preceded him—is Swartz's legacy. It is also his challenge to the world he left behind.

American intellectual-property statutes are rooted historically in their framers' disdain for the demos, and a dictatorial notion of culture. These laws, which are today weighted wholly in favor of producers, support and sustain the "push marketing" model, in which art and scholarship are generated by the creative elite and dispensed to a grateful public. This model implies an inherent separation between creators and consumers, in which the two parties are brought together only at the point of sale.

This mind-set conflicts with the way that people experience culture today. Internet users can easily opt to choose their own cultural adventures, and the creator of any given piece of content cannot effectively dictate the terms of its use. The value of authorial copyright has been diminished in the digital era, and just as nineteenth-century British authors responded to American reprinting practices by crying piracy and seeking legislative redress, contemporary culture-merchants use the language of morality to justify existing laws and lobby for new ones. But property holders are but one party to the social contract that is supposed to govern our polity, and their interests are not the only ones that matter. There is a middle ground between functionally eternal copyright and wholesale anarcho-syndicalism.

Intellectual-property policies are mutable social relationships among consumers, producers, and the state, and emerging societies, such as antebellum America or contemporary China, can find great national benefit in laws that promote the speedy circulation of knowledge and culture. ("It is only by the enlightening and education

of the people, that we can expect our institutions to hold together,"
American legislators who opposed international copyright told Captain Marryat in 1837.) As Lisa Rein indicated in 2002, the Internet
itself resembles an emerging, albeit stateless, society. Its users are
responsible for its welfare, and its welfare is inherently important,
because the Internet is the apotheosis of the cultural brain. The library of the future is here, and the value that we get out of it directly
correlates with the value we put into it. We can choose to feed the
cultural brain, or we can choose to starve it. The salient point is that
it is a *choice*, and it is ours to make.

The Internet is increasingly choked with various closed systems
that see their users as little more than sentient credit cards: not
just the restricted-access databases that Swartz decried, but with
smartphone applications and social networking services that cir-
cumscribe the digital commons. The copyright term extensions that
were granted by Congress in 1998 will begin to expire in 2018.
Soon after that, if history is any guide, the general copyright statute will be due for revision. Soon after *that*, if history is any guide,
copyright stakeholders will gather in private to draft a statute that
protects and advances their financial interests. Information wants to
be expensive. Information wants to be free. Today, with the developed world linked by digital networks that have not entirely fulfilled
their transformative promise, Stewart Brand's paradox seems more
relevant and more frustrating than ever.

Aaron Swartz spent his life caught in this paradox, and while he
didn't quite succeed in disentangling it, he at least called attention to
the fact that it exists. Three years on, the story of his life and death
serves as a necessary reminder that there is a fundamental disconnect
between our laws and our habits, between the way we are supposed
to conduct ourselves online and the way we actually do. But the gap
between the real and the ideal can be bridged only if we make the
active choice to extend ourselves. We can amend bad laws. We can
promote better modes of culture. We can choose to create systems

that can be opened without breaking; that tolerate deviance without collapsing; that regard the unfamiliar not as a threat, but as an opportunity.

When he lived in Massachusetts, Swartz regularly competed in the MIT Mystery Hunt.[51] Exactly one week after Swartz died, on the day the 2013 Mystery Hunt began, his old team hosted a memorial event in the lobby of the MIT Media Lab. A large banner was spread out on a table, on which friends and admirers inscribed personalized messages: funny memories, words of condolence. Near the end of the night, a slender boy in a plain sweatshirt who looked too young to be there came over to the table. He uncapped a marker. He wrote, simply, "We will continue."

ACKNOWLEDGMENTS

I want to thank Josh Levin for putting me on this story in the first place, and for his aid and friendship over the years. Thanks to Jonathan L. Fischer, David Plotz, John Swansburg, Julia Turner, and everyone at *Slate* for their patience and support. I'm grateful to Paul Whitlatch for acquiring the book, Brant Rumble for nurturing it, and Colin Harrison for seeing it home. Steve Boldt, Katrina Diaz, Kate Lloyd, and Elisa Rivlin provided invaluable guidance during the prepublication process. Jeff Greggs is a fine editor and a good friend. Thanks to my literary agent, Todd Shuster, for his wise counsel. Thanks to the University of Illinois Archives and its staffers, especially Cara Bertram. Thanks to all of my former colleagues at the *Columbia Journalism Review*. Thanks to Dean from Delta because I promised I would thank him in this space; I only wish I remembered his surname. I bet Jonathan Eck would get a kick out of seeing his name in a book. Many of my friends and colleagues offered their support while this book was under construction, but I am particularly grateful to Kelly Buttermore, Phil Campbell, Matt Demblowski, Noah Doyle, Sam Eifling, Mike Hoyt, Lauren Kirchner, Daniel Luzer, Matt B. Weir, and Joshua Young. I would like to thank my sister, Molly Peters. I would like to thank my parents, Randall and Janet Peters. Most of all, I would like to thank Alexa Mills, who knows why.

NOTES

INTRODUCTION: THE BAD THING

Parts of this chapter previously appeared, in slightly different format, in Justin Peters, "The Idealist," *Slate*, February 7, 2013, http://www.slate.com /articles/technology/technology/2013/02/aaron_swartz_he_wanted_to _save_the_world_why_couldn_t_he_save_himself.html. Reprinted with permission.

1 Interview with Taren Stinebrickner-Kauffman, January 2013.

2 Quinn Norton, "Life Inside the Aaron Swartz Investigation," *Atlantic*, March 3, 2013, http://www.theatlantic.com/technology/archive/2013/03 /life-inside-the-aaron-swartz-investigation/273654/.

3 Abelson et al., *Report to the President*, 39 (hereafter referred to as *MIT Report*). Accounts of this first plea offer differ. The US Attorney's Office told the MIT panel that Swartz was offered a three-month prison sentence; Swartz's former attorney, Andrew Good, said that the government offered him a thirteen-month sentence.

4 Norton, "Life Inside the Aaron Swartz Investigation."

5 Aaron Swartz, "HOWTO: Lose weight," *Raw Thought*, March 1, 2010, http://www.aaronsw.com/weblog/loseweight.

6 Danny O'Brien, "Teenager in a Million," *Sunday Times*, April 29, 2001, http://www.aaronsw.com/2002/teenagerInAMillion.

7 For the Reddit sale price, see Christine Lagorio-Chafkin, "How Alexis Ohanian Built a Front Page of the Internet," *Inc.*, May 30, 2012, http://

www.inc.com/magazine/201206/christine-lagorio/alexis-ohanian-reddit-how-i-did-it.html.

8 Norton, "Life Inside the Aaron Swartz Investigation."

9 Aaron Swartz, "Copyright Terrorism," *Aaron Swartz: The Weblog*, May 22, 2002, http://www.aaronsw.com/weblog/000277.

10 Aaron Swartz, "Guerilla Open Access Manifesto," Archive.org, https://archive.org/stream/GuerillaOpenAccessManifesto/Goamjuly2008_djvu.txt.

11 US Attorney's Office, District of Massachusetts, "Alleged Hacker Charged with Stealing Over Four Million Documents from MIT Network," news release, July 19, 2011, http://en.wikisource.org/wiki/Alleged_Hacker_Charged_with_Stealing_Over_Four_Million_Documents_from_MIT_Network.

12 John Summers and George Scialabba, "John Summers and George Scialabba: Statement in Support of Aaron Swartz," *Guernica*, July 23, 2011, https://www.guernicamag.com/daily/john_summers_and_george_sciala/.

13 Stinebrickner-Kauffman, interview.

14 Aaron Swartz, "Neurosis #9," *Raw Thought*, February 7, 2007, http://www.aaronsw.com/weblog/imposition.

15 Aaron Swartz, "Think Bigger: A Generalist Manifesto," *Raw Thought*, December 14, 2006, http://www.aaronsw.com/weblog/thinkbigger.

16 Aaron Swartz, "How to Save the World, Part 1," *Raw Thought*, July 28, 2011, http://www.aaronsw.com/weblog/save1.

17 Aaron Swartz, "2010 Review of Books," *Raw Thought*, January 3, 2011, http://www.aaronsw.com/weblog/books2010.

18 Seth David Schoen, "How I Knew Aaron," *Remember Aaron Swartz*, last modified January 16, 2013, http://www.rememberaaronsw.com/memories/How-I-Knew-Aaron.html.

19 Aaron Swartz, "Dave Winer said . . . ," *Aaron Swartz: The Weblog*, June 28, 2003, http://www.aaronsw.com/weblog/000988.

20 Aaron Swartz, "Bits are not a bug," *Not a Bug*, http://bits.are.notabug.com/.

21 Aaron Swartz, "In Offense of Classical Music," *Aaron Swartz: The Weblog*, June 20, 2006, http://www.aaronsw.com/weblog/classicalmusic.

22 Aaron Swartz, "The Anti-Suit Movement," *Raw Thought*, March 16, 2010, http://www.aaronsw.com/weblog/antisuit.

23 Interview with Wes Felter, February 2013.

24 Aaron Swartz, "Eat and Code," *Aaron Swartz: The Weblog*, August 2, 2005, http://www.aaronsw.com/weblog/eatandcode.

25 Interview with Ben Wikler, January 2013.

26 Stinebrickner-Kauffman, interview.

27 Ibid.

28 Interview with Nathan Woodhull, February 2013.

29 Stinebrickner-Kauffman, interview.

30 Ibid.

31 Ibid.

32 Brand, *Media Lab*, 202. Brand first used this phrase at a computing conference in 1984.

33 Ibid.

34 Rob Kling and Roberta Lamb, "Analyzing Alternate Visions of Electronic Publishing and Digital Libraries," in *Scholarly Publishing*, eds. Peek and Newby, 27.

1. NOAH WEBSTER AND THE MOVEMENT FOR COPYRIGHT IN AMERICA

1 Movable type had existed in Asia for four hundred years prior.

2 Blagden, *Stationers' Company*, 19.

3 Birrell, *Seven Lectures*, 22–23.

4 William Patry, "England and the Statute of Anne," *Copyright Law and Practice*, http://digital-law-online.info/patry/patry2.html.

5 Birrell, *Seven Lectures*, 94.

6 John L. Brooke, "Print and Politics," in Gross and Kelley, *Extensive Republic*, 180.

7 In 1671, Sir William Berkeley, the longtime colonial governor of Virginia, spoke for many when he wrote, "I thank God we have no free schools nor printing, and I hope we shall not have these hundred years; for learning has brought disobedience, and heresy, and sects into the world, and printing has divulged them; and libels against the best government. God keep us from both!"

8 Webster to William Leete Stone, December 21, 1837, in Webster, *Letters*, 511.

9 Ford, *Notes*, 1:16–31.

10 Kendall, *Forgotten Founding Father*, 50.

11 Ford, *Notes*, 1:38.

12 He eventually studied law on his own and in 1781 was admitted to the Connecticut bar.

13 Noah Webster, "Introduction to 'Blue-Back Speller,'" in *Autobiographies of Noah Webster*, 75.

14 Webster to Joel Barlow, November 12, 1807, in Ford, *Notes*, 2:31.

15 Webster to John Canfield, January 6, 1783, in Ford, *Notes*, 1:58.

16 Micklethwait, *Noah Webster and the American Dictionary*, 4.

17 Scudder, *Noah Webster*, 152.

18 Tim Cassedy, "'A Dictionary Which We Do Not Want': Defining America Against Noah Webster, 1783–1810," *William and Mary Quarterly* 71, no. 2 (April 2014): 229.

19 Noah Webster, "On the Education of Youth in America," in *Collection of Essays*, 24.

20 Charvat, *Profession of Authorship*, 18–34.

21 Charvat, *Literary Publishing*, 42.

22 Wolcott to Noah Webster, September 19, 1807, in Ford, *Notes*, 2:27.

23 James N. Green, "The Rise of Book Publishing," in Gross and Kelley, *Extensive Republic*, 78.

24 Letter from Joel Barlow to the Continental Congress, January 10, 1783, in *Primary Sources on Copyright (1450–1900)*, eds. L. Bently and M. Kretschmer, https://www.copyrighthistory.org.

25 Patry, *Moral Panics*, 192.

26 Williams, "Significance of the Printed Word," 43.

27 Noah Webster, "Origin of the Copy-Right Laws in the United States," in Webster, *Collection of Papers*, 173.

28 Ibid., 174.

29 It is unclear how much credit, if any, Webster deserves for the Connecticut law; as David Micklethwait has demonstrated, Connecticut was already considering a statewide copyright law by the time the state legislature received Webster's appeal.

30 Noah Webster, "Memoir," in *Autobiographies of Noah Webster*, 143.

31 Noah Webster, "Diary," in *Autobiographies of Noah Webster*, 222.

32 Smith, *Colonial Days*, 284.

33 Ford, *Notes*, 1:103.

34 Webster, "Diary," in *Autobiographies of Noah Webster*, 219.

35 When George Washington married Martha Custis in 1759, Custis already had two children from a previous marriage. The Washingtons had no children of their own.

36 Webster, "Memoir," in *Autobiographies of Noah Webster*, 143–44.

37 Noah Webster, "The News-Boy's Address to His Customers," in *Poems by Noah Webster*.

38 Oren Bracha, "The Adventures of the Statute of Anne in the Land of Unlimited Possibilities: The Life of a Legal Transplant," *Berkeley Technology Law Journal* 25, no. 1427 (2010): 1453.

39 William Patry, "The First Copyright Act," *Copyright Law and Practice*, http://digital-law-online.info/patry/patry5.html.

40 Ibid.

41 Meredith L. McGill, "Copyright," in Gross and Kelley, *Extensive Republic*, 199.

42 Richard R. John, "Expanding the Realm of Communications," in Gross and Kelley, *Extensive Republic*, 216.

43 Noah Webster, "The Editor's Address to the Public," in *American Minerva*, December 9, 1793.

44 Lee, *Daily Newspaper in America*, 711.

45 Coll, "Noah Webster," 174.

46 Ibid., 190.

47 Webster to Oliver Wolcott Jr., April 13, 1801, in Ford, *Notes*, 1:530.

48 Webster to Joel Barlow, November 16, 1798, in Webster, *Letters*, 192.

49 Webster to Benjamin Rush, December 15, 1800, in Webster, *Letters*, 228.

50 Cobbett, *Porcupine's Works*, 51.

51 Webster to Joel Barlow, November 12, 1807, in Ford, *Notes*, 2:32.

52 Webster to Oliver Wolcott Jr., June 18, 1807, in Snyder, *Defining Noah Webster*, 322.

53 Webster to David Ramsay, October 1807, in Webster, *Letters*, 291.

54 Ibid., 287.

55 Snyder, *Defining Noah Webster*, 323.

56 Ford, *Notes*, 2:116.

57 John, "Expanding the Realm," 216.

58 Andie Tucher, "Newspapers and Periodicals," in Gross and Kelley, *Extensive Republic*, 395.

59 Webster to Thomas Dawes, July 25, 1809, in Webster, *Letters*, 318–19.

60 Webster to John Jay, June 9, 1813, in Ford, *Notes*, 2:121.

61 Webster to John Pickering, December 1816, in Webster, *Letters*, 371.

62 Webster to Stephen van Rensselaer, November 5, 1821, in Webster, *Letters*, 406.

63 Webster to Samuel Latham Mitchill, December 12, 1823, in Webster, *Letters*, 411.

64 Ford, *Notes*, 2:293.

65 Ibid., 2:304–5.

66 Abraham Webster to Noah Webster, January 26, 1830, in Ford, *Notes*, 2:313.

67 Unger, *Noah Webster*, 307.

68 See Micklethwait, *Noah Webster and the American Dictionary*, for more on this.

69 Webster, "Origin of the Copy-Right Laws," 176.

70 Webster to William Chauncey Fowler, January 26, 1831, in Webster, *Letters*, 425.

71 Ibid.

72 See Catherine Seville, "The Statute of Anne: Rhetoric and Reception in the Nineteenth Century," *Houston Law Review* 47, no. 4 (2010): 819–75.

73 Meredith L. McGill, "Copyright," in Casper et al., *Industrial Book*, 159.

74 Webster to Rebecca Greenleaf Webster, February 7, 1831, in Ford, *Notes*, 2:326.

75 Webster to William Chauncey Fowler, January 29, 1831, in Webster, *Letters*, 425.

76 David Leverenz, "Men Writing in the Early Republic," in Gross and Kelley, *Extensive Republic*, 356.

2. A TAX ON KNOWLEDGE

1 Hannay, *Life of Frederick Marryat*, 77. The historical price conversion, roughly accurate as of July 2015, was performed using the Composite Price Index compiled by the United Kingdom's Office for National Statistics, and the equation proposed by Jim O'Donoghue, Louise Golding, and Grahame Allen in their paper "Consumer Price Inflation Since 1750," *Economic Trends* 604 (March 2004).

2 Ibid., 76–77.

3 Harper, *House of Harper*, 89–90.

4 Earl L. Bradsher, "Book Publishers and Publishing," in Trent et al., *The Cambridge History of American Literature*, 3:541 (half a million copies); and Petition of British Authors, Washington, DC (1837), *Primary Sources on Copyright*, eds. Bently and Kretschmer.

5 Editors' Table, *Knickerbocker* 15, no. 6 (June 1840): 529.

6 Dickens to Henry Austin, May 1, 1842, in *Letters of Charles Dickens*, 1:83.

7 Arno L. Bader, "Captain Marryat and the American Pirates," in *Library*, 1935, 328.

8 Hannay, *Life of Frederick Marryat*, 59.

9 The Locofocos were a faction of libertarian Democrats who enjoyed a brief period of national influence in the 1830s and 1840s.

10 Marryat, *Second Series*, 76.

11 Ibid., 70.

12 Marryat, *Life and Letters*, 2:19.

13 The American vessel was the steamboat *Caroline*, which had been aiding Canadian rebels who were resisting British rule. Loyalist forces captured the *Caroline* and sent it over Niagara Falls.

14 Marryat, *Diary in America*, 9.

15 Marryat, *Second Series*, 65.

16 McGrane, *Panic of 1837*, 1.

17 Dickens to John Forster, May 3, 1842, in Forster, *Life of Charles Dickens*, 1:409.

18 Marryat, *Second Series*, 64.

19 Putnam, *The Tourist in Europe*, 172.

20 Marryat, *Second Series*, 77.

21 Petition of British Authors, Washington, DC (1837), *Primary Sources on Copyright*, eds. Bently and Kretschmer.

22 Marryat, *Second Series*, 71.

23 *Debates in Congress*, 670.

24 Ibid., 671.

25 Bishop, "Struggle for International Copyright," 105–6.

26 Lee, *Daily Newspaper in America*, 718.

27 McLeod, *Pranksters*, 115–17.

28 Johnson, "The People's Book," 32.

29 Louise Stevenson, "Homes, Books, and Reading," in Casper et al., *Industrial Book*, 327.

30 Marryat, *Second Series*, 65.

31 Solberg, *International Copyright*, 10.

32 Quoted in Johnson, *Charles Dickens*, 1:367.

33 Dickens to John Forster, May 3, 1842, in Forster, *Life of Charles Dickens*, 1:409.

34 Mathews, *Various Writings*, 364.

35 "The International Copyright Law, and Mr. Dickens," *Arcturus* 3, no. 16 (March 1842): 247.

36 Miller, *Raven and the Whale*, 80.

37 Madeleine B. Stern, "Introduction," in *Publishers for Mass Entertainment*, ix.

38 Mathews, *Various Writings*, 363.

39 George Haven Putnam, "International Copyright.—VI," *Publishers' Weekly*, March 22, 1879, 351.

40 Weeks, *History of Paper-Manufacturing*, 287.

41 Ibid., 294.

42 Foner, *Great Labor Uprising*, 13.

43 Madeleine B. Stern, "Popular Books in the Midwest and Far West," *Getting the Books Out*, ed. Michael Hackenberg, 83.

44 Lawrence Parke Murphy, "Beadle & Co," in Stern, *Publishers for Mass Entertainment*, 49.

45 "Book-A-Day Lovell Set Many Records," *Montreal Gazette*, April 23, 1932.

46 Lovell, "The Canadian Incursion," *Publishers' Weekly*, April 19, 1879, 471.

47 Putnam, *Memories*, 169.

48 Ibid., 170.

49 Golway, *Machine Made*, 156–57.

50 "Our 'Moral Pirates,'" *Publishers' Weekly*, June 25, 1881, 693.

51 Advertisement in *Publishers' Weekly*, June 26, 1880, 666.

52 Ibid.

53 Putnam, *Memories*, 366.

54 Ibid., 367.

55 Putnam, "International Copyright.—VI."

56 George Haven Putnam, "International Copyright.—IV," *Publishers' Weekly*, March 8, 1879, 284.

57 "The Copyright Question—Opinions of Publishers and Authors—VII," *Publishers' Weekly*, April 19, 1879, 469.

58 Gilder to James Bryce, May 8, 1883, in Gilder, *Letters*, 117.

59 *Revision of Copyright Laws: Hearings Before the Committees on Patents of the Senate and the House of Representatives on Pending Bills to Amend and Consolidate the Acts Respecting Copyright*, 60th Cong. (1908) (statement of William Allen Jenner, March 27, 1908).

60 Putnam, *Memories*, 374.

61 Ibid., 376.

62 Ibid., 380.

63 Van Dyke, *National Sin*, 9.

64 Gardiner G. Hubbard, "International Copyright," *Science* 7, no. 158 (February 12, 1886): 137.

65 Grover Cleveland, "Second Annual Message (First Term)," December 6, 1886. Put online by Gerhard Peters and John T. Woolley, *The American Presidency Project*, http://www.presidency.ucsb.edu/ws/?pid=29527.

66 Gilder to Henry Adams, January 8, 1889, in Gilder, *Letters*, 202.

67 *Reports of Committees of the House of Representatives*, 7:25.

68 Twain, *Autobiography*, 2:319.

69 Johnson, *Remembered Yesterdays*, 246.

70 Ibid., 254.

71 Ibid., 257.

72 22 Cong. Rec. S3894 (March 3, 1891).

73 Ibid., S3905.

74 Johnson, *Remembered Yesterdays*, 258.

75 Ibid.

76 Ibid., 259.

77 Ibid.

78 Stedman and Gould, *Life and Letters*, 2:417.

79 Wilson, *Labor of Words*, 2.

80 Henry Holt, "The Commercialization of Literature," *Atlantic Monthly* 96 (November 1905): 588.

81 Wilson, *Labor of Words*, 88.

3. A COPYRIGHT OF THE FUTURE, A LIBRARY OF THE FUTURE

1 *Report of the Trustees of the Public Library of the City of Boston*, 15.

2 City of Boston Statistics Department, *Boston Statistics, 1922*, 42.

3 Herbert Putnam, "The Great Libraries of the United States," *Forum* 19 (June 1895): 492.

4 Quoted in Jones, "Constructing the Universal Library," 130.

5 Putnam, "Great Libraries."

6 Cole, *For Congress and the Nation*, 54.

7 Lee, *Daily Newspaper in America*, 715–17.

8 Herbert Putnam, "The Relation of Free Public Libraries to the Community," *North American Review* 166, no. 499 (June 1898): 667.

9 Carl F. Kaestle and Janice A. Radway, "A Framework for the History of Publishing and Reading in the United States, 1880–1940," in *Print in Motion*, eds. Kaestle and Radway, 14.

10 Wilson, *Labor of Words*, 3.

11 Putnam, "Relation of Free Public Libraries."

12 Library of Congress, *Herbert Putnam*, 9.

13 Nichols, *1883 Triennial*, 65.

14 Jane Aikin, "Referred to the Librarian, with Power to Act: Herbert Putnam and the Boston Public Library," in *Winsor, Dewey, and Putnam*, ed. Davis et al., 30.

15 Quoted in Knowlton, *Herbert Putnam: A 1903 Trip to Europe*, 47.

16 "The New Librarian," *Washington Evening Star*, April 5, 1899.

17 Putnam, "Relation of Free Public Libraries."

18 Cole, *For Congress and the Nation*, 75.

19 Brylawski and Goldman, *Legislative History of the 1909 Copyright Act*, 2:199.

20 Ibid., 1:171.

21 Ibid., 2:167.

22 Ibid., 1:3.

23 Ibid., 4J:120.

24 Ibid., 4H:24.

25 Ibid.

26 Ibid., 4J:310.

27 Ibid., 4J:321–33.

28 Ibid., 4H:141.

29 Ibid., 4J:33.

30 Ibid., 4H:98.

31 Ibid., 4H:115.

32 Ibid., 4J:315.

33 Ibid., 4J:33.

34 Ibid., 4H:3–4.

35 Ibid., 4H:77.

36 Quoted in Edward N. Waters, "Herbert Putnam: The Tallest Little Man in the World," *Quarterly Journal of the Library of Congress* 33, no. 2 (April 1976): 160. In 1904, Columbian College was rechristened the George Washington University.

37 Herbert Putnam, foreword to Koch, *War Service*, 6.

38 Quoted in Waters, "Herbert Putnam," 163.

39 Knowlton, *A 1903 Trip to Europe*, 13.

40 Ibid., 50.

41 Ibid., 171.

42 Matt McDade, "Man of Books, at 92, Places Listening Ahead of Reading," *Washington Post and Times Herald*, April 5, 1954.

43 In June 1967, Flushing Meadow Park was renamed Flushing Meadows–Corona Park.

44 "Dupont . . . at nywf64.com," *NYWF64.com*, http://www.nywf64.com/dupont01.shtml.

45 "Formica . . . at nywf64.com," *NYWF64.com*, http://www.nywf64.com/formica01.shtml

46 "United States . . . at nywf64.com," *NYWF64.com*, http://www.nywf64.com/unista09.shtml.

47 Putnam, "Great Libraries."

48 Overhage and Harman, *INTREX*, 145.

49 Vannevar Bush, "As We May Think," in Kochen, *Growth of Knowledge*, 32.

50 Overhage and Harman, *INTREX*, 144–45.

51 Grossman, *Omnibus Copyright Revision Legislative History*, 5:28.

52 Ibid., vol. 3, part 2, 97.

53 Ibid., vol. 3, part 3, 177.

54 Ibid., vol. 3, part 3, 154.

55 Ibid., 5:228–29.

56 Ibid., vol. 3, part 3, 100.

57 Ibid., vol. 3, part 2, 100.

58 Ibid., 5:197.

59 "School Mimeography and Copyright Law," *Chicago Tribune*, August 18, 1964.

60 Grossman, *Omnibus Copyright Revision Legislative History*, 5:63–64.

61 McSherry, *Who Owns Academic Work?*, 163–64. McSherry makes this distinction in the context of patent law, but I think it's relevant to copyright law as well.

62 Watson Davis, "The Universal Brain: Is Centralized Storage and Retrieval of All Knowledge Possible, Feasible, or Desirable?," in Kochen, *Growth of Knowledge*, 62.

63 Colin Burke, "A Rough Road to the Information Highway. Project Intrex: A View from the CLR Archives," *Information Processing & Management* 32, no. 1 (1996): 30.

64 Ibid.

4. THE INFINITE LIBRARIAN

1 Michael Stern Hart Papers, 1964–2010, University of Illinois Archives, Box 1, Folder "Essays" (hereafter cited as Hart Papers).

2 Ibid., Folder "College Material."

3 Ibid., Box 3, Folder "Correspondence 1966–67."

4 Ibid., Box 1, Folder "Essays."

5 Ibid.

6 Ibid., Box 3, Folder "Correspondence 1966–67."

7 Michael Hart, "Educate Mem," January 22, 2003, http://hart.pglaf.org /educate.mem.txt. They accepted him anyway.

8 Hart, "Educate Mem." On the same day, he also threatened to sue one of his teachers if she blocked his admittance into an advanced math and science program.

9 Hart Papers, Box 1, Folder "Essays."

10 Ibid.

11 Ibid., Box 3, Folder "Correspondence 1966–67."

12 Ibid.

13 Ibid.

14 Hart, "Educate Mem."

15 Hart Papers, Box 1, Folder "Life Review 1990."

16 Ibid., Folder "Essays."

17 Ibid.

18 Ibid.

19 "Letters of Recommendation for Michael S. Hart," https://web.archive.org/web/20110612153429/http://promo.net/hart/letters.html.

20 "SDS: The Sigma Family," ComputerHistory.org, http://archive.computerhistory.org/resources/text/SDS/SDS.Sigma.1967.102646100.pdf.

21 *Oxford English Dictionary*.

22 Louis T. Milic, "The Next Step," *Computers and the Humanities* 1, no. 1 (September 1966): 4.

23 "History of Project Gutenberg," *Project Gutenberg News*, http://www.gutenbergnews.org/about/history-of-project-gutenberg/.

24 Though Michael Hart claimed that the Xerox Sigma V was connected to the ARPANET when he first encountered it, records indicate that the University of Illinois did not become an official ARPANET node until later in 1971. It is likely that the network he remembered was a local one comprised of other campus computers.

25 Hart Papers, Box 16.

26 Hart Papers, Box 7, Folder "Early Written History of P.G. c. 1980."

27 Michael Hart to Book People mailing list, January 13, 2006, http://onlinebooks.library.upenn.edu/webbin/bparchive?year=2006&post=2006-01-13,13.

28 "History of Project Gutenberg," *Project Gutenberg News*.

29 Hart Papers, Box 9, Folder "Hearing."

30 Ibid., Box 7, Folder "Early Written History of P.G. c. 1980."

31 "History of Project Gutenberg," *Project Gutenberg News*.

32 The thirty-seven-thousand-plus number comes from Rebecca J. Rosen, "The Legacy of Project Gutenberg Founder, Michael S. Hart," *Atlantic*, September 8, 2011, http://www.theatlantic.com/technology/archive/2011/09/the-legacy-of-project-gutenberg-founder-michael-s-hart/244787/.

33 Michael Hart, "An Introduction to Michael S. Hart," January 1, 2006, http://hart.pglaf.org/intro.me.txt.

34 Michael Hart to Book People mailing list, January 12, 2006, http://onlinebooks.library.upenn.edu/webbin/bparchive?year=2006&post=2006-01-12,3.

35 Hart Papers, Box 1, Folder "Framed Aphorisms." Hart got the Blake quotation wrong. The actual line, from Blake's *Jerusalem: The Emanation of the Giant Albion*, reads, "I must Create a System, or be enslav'd by another Mans."

36 Hart to Book People mailing list, January 12, 2006.

37 Hart Papers, Box 6, Folder "Brainstorming Names for P.G. c. 1971."

38 Putnam, "Great Libraries."

39 Michael Hart, "The Cult of the Amateur," http://hart.pglaf.org/cult.of.the.amateur.txt.

40 Hart Papers, Box 1, Folder "Essays."

41 Hafner and Lyon, *Where Wizards Stay Up Late*, 34.

42 Licklider, *Libraries of the Future*, 6.

43 Williams, *Free as in Freedom (2.0)*, 78.

44 20 GOTO 10.

45 Williams, *Free as in Freedom (2.0)*, 54–55.

46 Hart Papers, Box 1, Folder "Journals, folder 1 of 2, 1979, 2000, undated."

47 Richard Stallman, "The GNU Project," www.gnu.org, http://www.gnu.org/gnu/thegnuproject.html.

48 Ibid.

49 Richard Stallman, "Copyleft: Pragmatic Idealism," www.gnu.org, https://www.gnu.org/philosophy/pragmatic.html.

50 Michael Hart, "Introduction to Michael Hart's blog," http://hart.pglaf.org/myblog.int.txt.

51 Hart Papers, Box 7, Folder "Project Gutenberg—Newsletters 1991–92."

52 Ibid., Box 1, Folder "Essays."

53 Ibid., Box 3, Folder "Correspondence 1974–1985."

54 Ibid., Box 9, Folder "Hymen Hart and Alice Woodby, c. 1975."

55 Ibid., Box 1, Folder "Life Review 1990."

56 Ibid., Box 8, Folder "Duncan Research—Correspondence 1987–90."

57 Ibid.

58 Ibid., Folder "Geof Pawlicki Folder 1 of 2 1987–88, 1992."

59 Ibid., Box 1, Folder "Life Review 1990."

60 Ibid., Box 2, Folder "American Library Association Midwinter."

61 Gillies and Cailliau, *How the Web Was Born*, 209.

62 Hart Papers, Box 6, Folder "Project Gutenberg—Correspondence 1992, 2 of 2."

63 Ibid., Folder "Project Gutenberg—Correspondence 1993."

64 Ibid., Box 3, Folder "Correspondence 1995."

65 Ibid., Box 8, Folder "Electronic Networking 1991."

66 Elizabeth Weise, "Project Gutenberg Puts Great Literature on the Internet," Associated Press, 1995.

67 Michael Hart, *A Brief History of the Internet*, http://archive.org/stream /abriefhistoryoft00250gut/pg250.txt.

68 Ibid.

69 US Senate Committee on the Judiciary, *Copyright Term Extension Act of 1995*, 44.

70 Ibid., 57.

71 Ibid., 55.

72 Ibid., 41.

73 Ibid., 42.

74 Ibid., 71.

75 Michael Hart, "PG Newsletter March 1995," March 1, 1995, http://www .gutenbergnews.org/19950301/pg-monthly-newsletter-1995-03/.

76 Michael Hart, "Why I started [*sic*] My Blog," http://hart.pglaf.org/whyblog .txt.

77 Hart Papers, Box 6, Folder "Project Gutenberg—Correspondence 1985–89."

78 Ibid., Box 7, Folder "Letters of Support for Project Gutenberg, Folder 2 of 3."

79 Michael Hart, "Project Gutenberg #500," March 1996, http://www.ub .uni-dortmund.de/listen/inetbib/msg04134.html.

80 Hart Papers, Box 7, Folder "Letters of Support for Project Gutenberg, Folder 1 of 3, 1996."

81 Ibid.

82 Ibid., Folder "Letters of Support for Project Gutenberg, Folder 2 of 3."

83 Michael Hart, "PG Newsletter June 1996," June 9, 1996, http://www .gutenbergnews.org/19960609/pg-monthly-newsletter-1996-06-09/.

84 Michael Hart to Book People mailing list, September 4, 1997, http:// onlinebooks.library.upenn.edu/webbin/bparchive?year=1997&post=1997 -09-04$4.

5. THE CASE FOR THE PUBLIC DOMAIN

1 Andrea L. Foster, "A Bookworm's Battle," *Chronicle of Higher Education*, October 25, 2002, http://chronicle.com/article/A-Bookworm-s-Battle/10315.

2 Ibid.

3 "The Copyright Question—Opinions of Publishers and Authors—III," *Publishers' Weekly*, March 1, 1879, 262.

4 Eric Eldred, "Battle of the Books," last modified November 15, 1998, https://web.archive.org/web/20000817031559/http://www.eldritchpress.org/battle.html.

5 Ibid.

6 In 1998, Eldred wrote, "I thought it was all in the public domain, or at least thought there was a good argument for that. But I have been more aggressive than most in going after material published after 1922. If the work was published before 1989 and didn't have a copyright notice on it I considered it in the public domain." See "Battle of the Books."

7 Carl S. Kaplan, "Online Publisher Challenges Copyright Law," *New York Times*, January 15, 1999.

8 On the lobbying efforts by Disney and other media companies, see Darin Fonda, "Copyright's Crusader," *Boston Globe Magazine*, August 29, 1999.

9 Representative Bono (CA), "Sonny Bono Copyright Term Extension Act," *Congressional Record* 144:139 (October 7, 1998), H9952.

10 Representative McCollum (FL), "Sonny Bono Copyright Term Extension Act," *Congressional Record* 144:139 (October 7, 1998), H9951.

11 See *Universal Studios, Inc. v. Corley*, a case in which the Southern District of New York cited the DMCA to issue an injunction preventing Eric "Emmanuel Goldstein" Corley, publisher of the hacker magazine *2600*, from posting to his website a decryption program called DeCSS, and from linking to other websites that hosted the program. The Second Circuit, on appeal, upheld the district court's injunction.

12 Sandra A. Sellers, Software Publishers Association, to the House Subcommittee on Courts and Intellectual Property, Committee on the Judiciary, US House of Representatives, 105th Cong., 1st sess., September 11, 1997.

13 "Judge Dismisses Indictment Against MIT Computer Whiz," Reuters, December 29, 1994.

14 Marybeth Peters, to the House Subcommittee on Courts and Intellectual Property, Committee on the Judiciary, US House of Representatives, 105th Cong., 1st sess., September 11, 1997.

15 Eldred, "Battle of the Books."

16 Michael Hart to Book People mailing list, October 19, 1998, http://onlinebooks.library.upenn.edu/webbin/bparchive?year=1998&post=1998-10-22$5.

17 Michael Hart to Book People mailing list, October 19, 1998, http://onlinebooks.library.upenn.edu/webbin/bparchive?year=1998&post=1998-10-19$5.

18 Eric Eldred to Book People mailing list, October 19, 1998, http://onlinebooks
    .library.upenn.edu/webbin/bparchive?year=1998&post=1998-10-19$4.

19 Kaplan, "Online Publisher Challenges."

20 Steven Levy, "Lawrence Lessig's Supreme Showdown," *Wired*, October
    2002, http://archive.wired.com/wired/archive/10.10/lessig.html.

21 David Streitfeld, "The Cultural Anarchist vs. the Hollywood Police State,"
    Los Angeles Times, September 22, 2002, http://articles.latimes.com/2002
    /sep/22/magazine/tm-copyright38.

22 Naftali Bendavid, "Lawyer in Microsoft Case Cut Teeth at U. of C."
    Chicago Tribune, January 26, 1998, http://articles.chicagotribune.com
    /1998-01-26/news/9801260179_1_lawrence-lessig-judge-richard-posner
    -judge-thomas-penfield-jackson.

23 Lawrence Lessig, "How I Lost the Big One," *Legal Affairs*, March/April
    2004, http://www.legalaffairs.org/issues/March-April-2004/story_lessig
    _marapr04.msp.

24 Richard Poynder, "The Basement Interviews: Free Culture," *The Base-
    ment Interviews*, April 7, 2006, http://ia802307.us.archive.org/23/items/
    The_Basement_Interviews/Lawrence_Lessig_Interview.pdf.

25 Foster, "Bookworm's Battle."

26 Eldred, "Battle of the Books."

27 Lisa Rein, "Seth's Eldred Experience," *On Lisa Rein's Radar*, November 15,
    2002, http://www.onlisareinsradar.com/wp/seths-eldred-experience/.

28 Lisa Rein, "Aaron Swartz Camping Out at Eldred Oct 2002," Archive.org,
    https://archive.org/details/AaronSwartzEldredOct2002.

29 Aaron Swartz, "to the courthouse," *Aaron Swartz: The Weblog*, October 5,
    2002, http://www.aaronsw.com/weblog/000637.

30 Aaron Swartz, "Ahh, the first day of school," *Schoolyard Subversion*,
    August 29, 2000, http://web.archive.org/web/20010516224049/http://
    swartzfam.com/aaron/school/2000/08/29/.

31 Aaron Swartz, "Assorted Documents," 2002, http://www.aaronsw.com
    /2002/bizcard; and Swartz, "everyone looks like me," *Aaron Swartz: The
    Weblog*, July 22, 2002, http://www.aaronsw.com/weblog/000435.

32 Interview with Robert Swartz, January 2013.

33 Philipp Lenssen, "A Chat with Aaron Swartz," *Google Blogoscoped*,
    May 7, 2007, http://blogoscoped.com/archive/2007-05-07-n78.html.

34 To be clear, The Info Network was never nearly as popular as Wikipedia.

35 Lisa B. Song, "Highland Park Teen Is Finalist in Web Competition," *Chi-
    cago Tribune*, June 23, 2000, http://articles.chicagotribune.com/2000-06
    -23/news/0007290040_1_world-wide-web-arsdigita-boot-camp.

36 Zooko Wilcox-O'Hearn, "Part 1," *thoughts upon news of the death of Aaron Swartz*, February 6, 2013, http://zooko-on-aaronsw.blogspot.com /2013/02/part-1-by-zooko-wilcox-ohearn-written.html.

37 Interview with Dan Connolly, January 2013.

38 Gabe Beged-Dov to Discussion of XML News mailing list, July 3, 2000, https:// groups.yahoo.com/neo/groups/syndication/conversations/messages/241.

39 Aaron Swartz to Discussion of XML News mailing list, July 3, 2000, https:// groups.yahoo.com/neo/groups/syndication/conversations/topics/242.

40 Aaron Swartz, "I narrowly escaped . . . ," *Schoolyard Subversion*, August 16, 2000, http://web.archive.org/web/20010514192627/http://swartz fam.com/aaron/school/2000/08/16/.

41 Aaron Swartz, "The Weight of School," *Schoolyard Subversion*, October 8, 2000, http://web.archive.org/web/20010517235916/http://swartzfam.com /aaron/school/2000/10/08/.

42 Aaron Swartz, "Welcome to Unschooling," *Schoolyard Subversion*, April 5, 2001, http://web.archive.org/web/20010502005216/http:/swartzfam.com /aaron/school/2001/04/05/.

43 Robert Swartz, interview.

44 Tim Berners-Lee, James Hendler, and Ora Lassila, "The Semantic Web." *Scientific American*, May 17, 2001, http://www.cs.umd.edu/~golbeck /LBSC690/SemanticWeb.html.

45 Aaron Swartz, "I think there is a," *Aaron Swartz: The Weblog*, January 14, 2002, http://www.aaronsw.com/weblog/000111.

46 Felter, interview.

47 Wilcox-O'Hearn, "Part 1."

48 Eldred, "Battle of the Books."

49 Interview with Lisa Rein, January 2013.

50 Interview with Ben Adida, January 2013.

51 Rein, interview.

52 Aaron Swartz, "Emerging Technologies—Day 2," *Aaron Swartz: The Weblog*, May 15, 2002, http://www.aaronsw.com/weblog/000254.

53 Aaron Swartz, "May 13, 2002: Visiting Google," *Google Weblog*, May 13, 2002, http://google.blogspace.com/archives/000252.

54 Felter, interview.

55 Aaron Swartz, "Emerging Technologies—Day 3," *Aaron Swartz: The Weblog*, May 16, 2002, http://www.aaronsw.com/weblog/000255.

56 Ibid.

57 Eric Eldred to Book People mailing list, October 19, 1998, http://onlinebooks .library.upenn.edu/webbin/bparchive?year=1998&post=1998-10-19$4.

58 Recording Industry Association of America, "2002 Year-End Statistics," 2002, https://web.archive.org/web/20040601160725/http://www.riaa.com/news/marketingdata/pdf/year_end_2002.pdf.

59 *Music on the Internet: Is There an Upside to Downloading? Hearing before the Committee on the Judiciary*, 106th Cong., 2nd sess., July 11, 2000, http://www.gpo.gov/fdsys/pkg/CHRG-106shrg74728/pdf/CHRG-106shrg74728.pdf.

60 Ibid.

61 Brief for the RIAA as Amicus Curiae, 11, *Eldred v. Ashcroft*, 537 U.S. 186 (2003).

62 Brief for Dr. Seuss Enterprises, L.P., et al., as Amici Curiae, 9, *Eldred v. Ashcroft*, 537 U.S. 186 (2003).

63 Richard Koman, "Riding along with the Internet Bookmobile," *Salon*, October 9, 2002, http://www.salon.com/2002/10/09/bookmobile/.

64 Lawrence Lessig, "Brewster's Brilliance," *Lessig*, October 17, 2002, http://www.lessig.org/2002/10/brewsters-brilliance/.

65 Poynder, "The Basement Interviews."

66 Lawrence Lessig, "from the front line," October 13, 2002, http://www.lessig.org/2002/10/from-the-front-line/.

67 Aaron Swartz, "Arrgh, pirates," *Aaron Swartz: The Weblog*, February 04, 2002, http://www.aaronsw.com/weblog/000158.

68 Aaron Swartz, "LimeWire has gotten really good," *Aaron Swartz: The Weblog*, January 24, 2002, http://www.aaronsw.com/weblog/000143.

69 Aaron Swartz, "Trip Notes," *Aaron Swartz: The Weblog*, October 08, 2002, http://www.aaronsw.com/weblog/000647.

70 Aaron Swartz, "Mr. Swartz Goes to Washington," *Aaron Swartz: The Weblog*, October 10, 2002, http://www.aaronsw.com/weblog/000650.

71 Ibid.

72 Eric Eldred to Book People mailing list, October 14, 2002, http://onlinebooks.library.upenn.edu/webbin/bparchive?year=2002&post=2002-10-14,4.

73 Swartz, "Mr. Swartz Goes to Washington."

74 *Eldred v. Ashcroft*, 537 US 186 (2003).

75 Lisa Rein, "It's Over. We Lose," *On Lisa Rein's Radar*, January 15, 2003, http://www.onlisareinsradar.com/archives/000831.php#000831.

76 Cory Doctorow, "Supreme Court rules against Eldred, Alexandria burns," *Boing Boing*, January 15, 2003, http://boingboing.net/2003/01/15/supreme-court-rules.html.

77 Eric Eldred to Book People mailing list, January 15, 2003, http://onlinebooks.library.upenn.edu/webbin/bparchive?year=2003&post=2003-01-15,10.

78 Michael Hart to Book People mailing list, January 15, 2003, http://onlinebooks.library.upenn.edu/webbin/bparchive?year=2003&post=2003-01-15,7.

79 Lessig, *Free Culture*, 244–45.

80 Aaron Swartz, "Day of Mourning," *Aaron Swartz: The Weblog*, January 15, 2003, http://www.aaronsw.com/weblog/000806.

81 Swartz, "Mr. Swartz Goes to Washington."

6. "CO-OPT OR DESTROY"

1 Aaron Swartz, "Checking In," *Schoolyard Subversion*, December 23, 2001, http://web.archive.org/web/20020205111032/http:/swartzfam.com/aaron/school/.

2 Aaron Swartz, "Instant Message from LelandJr247," *Aaron Swartz: The Weblog*, December 11, 2003, http://www.aaronsw.com/weblog/001087.

3 Aaron Swartz, "Stanford: Day 1," *Aaron Swartz: The Weblog*, September 21, 2004, https://web.archive.org/web/20041009200559/http://www.aaronsw.com/weblog/001418.

4 Aaron Swartz, "Stanford: Day 3," *Aaron Swartz: The Weblog*, last modified June 3, 2005, http://www.aaronsw.com/weblog/001421.

5 Interview with Seth Schoen, January 2013.

6 Aaron Swartz, "Stanford: Day 58," *Aaron Swartz: The Weblog*, November 15, 2004, http://www.aaronsw.com/weblog/001480.

7 Wilcox-O'Hearn, "Part 1."

8 Aaron Swartz, "Home Again," *Aaron Swartz: The Weblog*, February 13, 2005, http://www.aaronsw.com/weblog/001558.

9 Aaron Swartz, "News Update," *Aaron Swartz: The Weblog*, February 17, 2003, http://www.aaronsw.com/weblog/000838.

10 Paul Graham, "What I Did This Summer," *PaulGraham.com*, October 2005, http://www.paulgraham.com/sfp.html.

11 Paul Graham, "Summer Founders Program," *PaulGraham.com*, March 2005, http://paulgraham.com/summerfounder.html.

12 Ibid.

13 Infogami, March 4, 2006, https://web.archive.org/web/20060323211212/http://infogami.com/.

14 Aaron Swartz, "infogami," *Infogami*, circa October 25, 2005, https://web.archive.org/web/20051025013124/http://infogami.com/. Swartz had two co-applicants: the Danish programmer Simon Carstensen and the British programmer and historian Sean B. Palmer.

15 Aaron Swartz, "SFP: Come see us," *Aaron Swartz: The Weblog*, April 16, 2005, http://www.aaronsw.com/weblog/001679.

16 Ibid.

17 Ibid.

18 Interview with Simon Carstensen, February 2013.

19 Aaron Swartz, "Introducing Infogami," *Infogami*, circa March 4, 2006, https://web.archive.org/web/20060304203354/http://infogami.com/blog/introduction.

20 Interview with Steve Huffman, January 2013.

21 Aaron Swartz, "Stanford: The Cynic Returns," *Aaron Swartz: The Weblog*, April 16, 2005, http://www.aaronsw.com/weblog/001680.

22 Aaron Swartz, "I Love the University," *Aaron Swartz: The Weblog*, July 26, 2006, http://www.aaronsw.com/weblog/visitingmit.

23 Aaron Swartz, "On Losing Weight," *Aaron Swartz: The Weblog*, July 26, 2006, http://www.aaronsw.com/weblog/losingweight.

24 Aaron Swartz, "A Night at the Coop," *Raw Thought*, October 24, 2006, http://www.aaronsw.com/weblog/coopnight.

25 Swartz, "Stanford: The Cynic Returns."

26 Aaron Swartz, "A Non-Programmer's Apology," *Aaron Swartz: The Weblog*, May 27, 2006, http://www.aaronsw.com/weblog/nonapology.

27 Aaron Swartz, "Stanford: Mr. Unincredible," *Aaron Swartz: The Weblog*, March 26, 2005, http://www.aaronsw.com/weblog/001629.

28 Rob Kling and Roberta Lamb, "Analyzing Alternate Visions of Electronic Publishing and Digital Libraries," in *Scholarly Publishing*, eds. Peek and Newby, 28.

29 Aaron Swartz, "Bits are not a bug," http://web.archive.org/web/20031229025933/http://bits.are.notabug.com/.

30 Aaron Swartz, "The Fountainhead by Ayn Rand," *Aaron Swartz: The Weblog*, February 3, 2002, http://www.aaronsw.com/weblog/000155.

31 Aaron Swartz, "Stanford: Private Meeting," *Aaron Swartz: The Weblog*, March 26, 2005, http://www.aaronsw.com/weblog/001645.

32 "iTunes Music Store. Facelift for a corrupt industry," *Downhill Battle*, http://www.downhillbattle.org/itunes/.

33 Aaron Swartz, "Of Washington and Worcester," *Aaron Swartz: The Weblog*, July 3, 2005, http://www.aaronsw.com/weblog/downhillbattle.

34 Swartz, "Stanford: Day 3."

35 Swartz, "Of Washington and Worcester."

36 Ibid.

37 George Anders, "Today I Learned: Reddit Could Be Worth $240 Million," *Forbes*, November 19, 2012, http://www.forbes.com/sites/georgeanders/2012/10/31/what-is-reddit-worth/.

38 Aaron Swartz, "The Aftermath," *Raw Thought*, November 1, 2006, http://www.aaronsw.com/weblog/theaftermath.

39 Ibid.

40 Aaron Swartz, "Business 'Ethics,'" *Raw Thought*, December 11, 2006, http://www.aaronsw.com/weblog/bizethics.

41 Aaron Swartz, "How to Get a Job Like Mine," *Aaron Swartz*, last updated August 18, 2008, https://aaronsw.jottit.com/howtoget.

42 Huffman, interview.

43 David Weinberger, "Reddit acquired," *Joho the Blog*, October 31, 2006, https://web.archive.org/web/20061102042749/http://www.hyperorg.com/blogger/mtarchive/reddit_acquired.html.

44 Mathew Ingram, "Reddit gets to Digg-ify Conde Nast," *MathewIngram.com*, October 31, 2006, http://www.mathewingram.com/work/2006/10/31/reddit-gets-to-digg-ify-conde-nast/.

45 Chris Slowe, a Harvard doctoral student who had joined Reddit in November 2005 as a part-time programmer, also profited from Not a Bug's sale.

46 Aaron Swartz, "The Afterparty," *Raw Thought*, November 2, 2006, http://www.aaronsw.com/weblog/theafterparty.

47 Aaron Swartz, "Office Space," *Raw Thought*, November 15, 2006, http://www.aaronsw.com/weblog/officespace.

48 Huffman, interview.

49 Aaron Swartz, "A Moment Before Dying," *Raw Thought*, January 18, 2007, http://www.aaronsw.com/weblog/dying. Swartz later edited the post to change the name of the "Aaron" character to "Alex."

50 Aaron Swartz, comment on "Reddit cofounder Aaron Swartz discusses how he was fired from Reddit," accessed on August 20, 2015, https://www.reddit.com/r/reddit.com/comments/1octb/reddit_cofounder_aaron_swartz_discusses_how_he/c1oezk.

51 Aaron Swartz, "Last Day of Summer Camp," *Raw Thought*, January 22, 2007, http://www.aaronsw.com/weblog/summercamp.

52 Ibid.

53 Aaron Swartz, "The Interrupt-Driven Life," *Raw Thought*, August 20, 2007, http://www.aaronsw.com/weblog/interruptdriven.

54 Aaron Swartz, "Everything Good Is Bad For You," *Raw Thought*, March 29, 2007, http://www.aaronsw.com/weblog/everythinggood.

55 Aaron Swartz, "Launch," *Raw Meat*, May 3, 2007, http://qblog.aaronsw .com/post/1483459/launch.

56 Aaron Swartz, "Announcing the Open Library," *Raw Thought*, July 16, 2007, http://www.aaronsw.com/weblog/openlibrary.

57 Becky Hogge, "Brewster Kahle," *New Statesman*, October 17, 2005, http:// www.newstatesman.com/node/151776.

58 Jones, "Constructing the Universal Library," 275.

59 Ibid., 276.

60 For more on the differing approaches of Open Library and Google Books, see Jones, "Constructing the Universal Library."

61 Heidi Benson, "A Man's Vision: World Library Online," *San Francisco Chronicle*, November 22, 2005, http://www.sfgate.com/news/article /A-MAN-S-VISION-WORLD-LIBRARY-ONLINE-Brewster-2593527 .php.

62 Jones, "Constructing the Universal Library," 282.

63 Swartz, "Announcing the Open Library."

64 Aaron Swartz, "Bubble City: Preface," *Raw Thought*, October 31, 2007, http://www.aaronsw.com/weblog/bubblecity.

65 Aaron Swartz, "Bubble City: Chapter 2," *Raw Thought*, November 2, 2007, http://www.aaronsw.com/weblog/bubblecity2.

66 Aaron Swartz, "Bubble City: Chapter 4," *Raw Thought*, November 6, 2007, http://www.aaronsw.com/weblog/bubblecity4.

67 Aaron Swartz, "Bubble City: Chapter 10," *Raw Thought*, November 19, 2007, http://www.aaronsw.com/weblog/bubblecity10.

68 Aaron Swartz, "Bubble City: Chapter 5," *Raw Thought*, November 6, 2007, http://www.aaronsw.com/weblog/bubblecity5.

69 Swartz, "Bubble City: Chapter 10."

70 Aaron Swartz, "Sick," *Raw Thought*, November 27, 2007, http://www .aaronsw.com/weblog/verysick.

71 Aaron Swartz, "Moving On," *Raw Thought*, June 16, 2008, http://www .aaronsw.com/weblog/movingon.

72 Aaron Swartz, "Banff," *Raw Thought*, March 16, 2008, http://www .aaronsw.com/weblog/banff.

7. GUERILLA OPEN ACCESS

1 *Eremo: Historical Memories.*

2 "St. Romuald," *Catholic News Agency*, http://www.catholicnewsagency .com/saint.php?n=510.

3 *Eremo: Historical Memories*, 19.

4 Ibid., 20.

5 Ibid., 61.

6 Ibid., 79.

7 Aaron Swartz, "Everybody Tells Me So," *Raw Thought*, November 3, 2006, http://www.aaronsw.com/weblog/everybodysays.

8 Aaron Swartz, "Aaron's Patented Demotivational Seminar," *Raw Thought*, March 27, 2007, http://www.aaronsw.com/weblog/demotivate.

9 Aaron Swartz, "The Book That Changed My Life," *Aaron Swartz: The Weblog*, May 15, 2006, http://www.aaronsw.com/weblog/epiphany.

10 Lawrence Lessig, "Required Reading: the next 10 years," *Lessig*, June 19, 2007, http://www.lessig.org/2007/06/required-reading-the-next-10-y-1/.

11 Ibid.

12 "Open Government Working Group," last updated October 22, 2007, https://public.resource.org/open_government_meeting.html; and Aaron Swartz, "announce theinfo.org," January 15, 2008, https://groups.google.com/forum/?hl=en#!topic/open-government/EzIgMgVHy1o.

13 Aaron Swartz, *TheInfo.org*, crawled on April 26, 2009, https://web.archive.org/web/20090426012606/http://theinfo.org/.

14 Aaron Swartz to Carl Malamud, December 31, 2007, https://public.resource.org/aaron/pub/msg00054.html.

15 Aaron Swartz to Open Government mailing list, March 8, 2008, https://public.resource.org/aaron/pub/msg00064.html.

16 Aaron Swartz, "Welcome, watchdog.net," *Raw Thought*, April 14, 2008, http://www.aaronsw.com/weblog/watchdog.

17 Song, "Highland Park Teen Is Finalist in Web Competition."

18 To give just one example, as of March 2015, an institutional print subscription to the journal *Applied Surface Science*, published by Reed Elsevier, cost $12,471.

19 Okerson and O'Donnell, *Scholarly Journals*, 105.

20 Ibid., 37.

21 "Budapest Open Access Initiative," February 14, 2002, http://www.budapestopenaccessinitiative.org/read.

22 All italicized quotes in this section come from the "Guerilla Open Access Manifesto," https://archive.org/stream/GuerillaOpenAccessManifesto/Goamjuly2008_djvu.txt.

23 Aaron Swartz, "OCLC on the Run," *Raw Thought*, November 15, 2008, http://www.aaronsw.com/weblog/oclreply.

24 Ibid.

25 Bess Sadler, "the free flow of information and ideas to present and future generations," *Solvitur ambulando*, January 15, 2013, http://www.ibiblio .org/bess/?p=285.

26 Malamud, *Exploring the Internet*, ix.

27 Carl Malamud, "The ITU Adopts a New Meta-Standard: Open Access," *Interoperability Report 5*, no. 12 (December 1991), https://public.resource .org/scribd/2556197.pdf.

28 Malamud, *Exploring the Internet*, 3.

29 Ibid., 4.

30 "Connexions," *Interoperability Report 5*, no. 12 (December 1991), 21.

31 Malamud, *Exploring the Internet*, ix.

32 Sharon Fisher, "ITU Standards Program to End," *Communications Week*, December 23, 1991.

33 Carl Malamud, "ITU Decision Turns Back the Clock," *Communications Week*, December 23, 1991, 14.

34 Malamud, *Exploring the Internet*, x.

35 Peter H. Lewis, "In 1996, a World's Fair to Be Held in Cyberspace," *New York Times*, March 14, 1995, http://www.nytimes.com/1995/03/14 /business/in-1996-a-world-s-fair-to-be-held-in-cyberspace.html.

36 Carl Malamud, "Lifting every voice," *St. Petersburg Times,* March 7, 1993.

37 Malamud, *World's Fair*, 94.

38 Ross Kerber, "What Cost Public Information?" *Washington Post*, June 21, 1993, http://www.washingtonpost.com/archive/business/1993/06/21 /what-cost-public-information-secs-new-data-system-prompts-debate-on -commercial-control-of-government-records/f493c24e-cb16-4c51-8b80 -f05a9637a6c0/.

39 Malamud, *World's Fair*, 94.

40 Ibid., 99.

41 Peter H. Lewis, "Internet Users Get Access to S.E.C. Filings Fee-Free," *New York Times*, January 17, 1994, http://www.nytimes.com/1994/01/17 /business/internet-users-get-access-to-sec-filings-fee-free.html.

42 Malamud, *World's Fair*, 95.

43 John Schwartz, "A Cybernaut Plans Software for Navigating TV," *New York Times*, December 24, 2001, http://www.nytimes.com/2001/12/24/ business/a-cybernaut-plans-software-for-navigating-tv.html.

44 "Public.Resource.Org—A 501(c)(3) Nonprofit Corporation," https:// public.resource.org/about/.

45 28 US Code § 1913—Courts of Appeals, https://www.law.cornell.edu /uscode/text/28/1913.

46 Stephen Schultze, "Electronic Public Access Fees and the United States Federal Courts' Budget: An Overview," http://www.openpacer.org/hogan/Schultze_Judiciary_Budgeting.pdf.

47 Stephen Schultze, "What Does It Cost to Provide Electronic Access to Court Records?," *Managing Miracles: Policy for the Network Society*, May 29, 2010, http://managingmiracles.blogspot.com/2010/05/what-is-electronic-public-access-to.html.

48 Ryan Singel, "Online Rebel Publishes Millions of Dollars in U.S. Court Records for Free," *Wired*, December 12, 2008.

49 Ibid.

50 "16 Frequently Asked Questions about Recycling Your PACER Documents," https://public.resource.org/uscourts.gov/recycling.html.

51 US Courts, "Pilot Project: Free Access to Federal Court Records at 16 Libraries," news release, November 8, 2007, https://web.archive.org/web/20071112035705/http://www.uscourts.gov/Press_Releases/libraries110807.html.

52 Carl Malamud in *The Internet's Own Boy* (2014), at 32:05.

53 "16 Frequently Asked Questions."

54 Carl Malamud to Aaron Swartz, September 4, 2008, 2:59 p.m., https://public.resource.org/aaron/pub/msg00180.html.

55 Aaron Swartz to Carl Malamud, September 4, 2008, 2:57 p.m., https://public.resource.org/aaron/pub/msg00179.html.

56 Aaron Swartz, "today's featured superhero: Carl Malamud," *Aaron Swartz: The Weblog*, June 16, 2002, http://www.aaronsw.com/weblog/000345.

57 Aaron Swartz, "Introducing web.resource.org," *Aaron Swartz: The Weblog*, July 2, 2002, http://www.aaronsw.com/weblog/000381.

58 Carl Malamud to Aaron Swartz, June 25, 2002, 8:13 p.m., https://public.resource.org/aaron/pub/msg00001.html.

59 US Courts, "Pilot Project: Free Access to Federal Court Records at 16 Libraries," news release, November 8, 2007, https://web.archive.org/web/20071206094450/http://www.uscourts.gov/Press_Releases/libraries110807.html.

60 Eric Hellman, "The Four Crimes of Aaron Swartz (#aaronswnyc part 2)," *go to hellman*, January 25, 2013, http://go-to-hellman.blogspot.com/2013/01/the-four-crimes-of-aaron-swartz.html.

61 Carl Malamud to Aaron Swartz, September 4, 2008, 7:40 p.m., https://public.resource.org/aaron/pub/msg00195.html.

62 Aaron Swartz to Carl Malamud, September 4, 2008, 10:41 p.m., https://public.resource.org/aaron/pub/msg00196.html.

63 Carl Malamud to Aaron Swartz, September 4, 2008, 7:43 p.m., https://public.resource.org/aaron/pub/msg00197.html.

64 Carl Malamud to Aaron Swartz, September 22, 2008, 7:26 p.m., https://public.resource.org/aaron/pub/msg00298.html.

65 Aaron Swartz to Carl Malamud, September 22, 2008, 10:37 p.m., https://public.resource.org/aaron/pub/msg00299.html.

66 Aaron Swartz, "Guerilla Open Access," *Raw Thought*, September 20, 2008, http://www.aaronsw.com/weblog/goa (page deleted).

67 "Software Freedom Day, Sept. 20th, 2008," *Free Software Foundation*, September 29, 2008, https://www.fsf.org/blogs/membership/sfd2008blog.

68 Aaron Swartz to Carl Malamud, October 1, 2008, 1:12 p.m., https://public.resource.org/aaron/pub/msg00327.html.

69 Carl Malamud to Aaron Swartz, September 30, 2008, 5:31 p.m., https://public.resource.org/aaron/pub/msg00319.html.

70 American Association of Law Libraries, *AALL Washington E-Bulletin*, October 31, 2008, http://www.aallnet.org/mm/Advocacy/aallwash/Washington-E-Bulletin/2008/ebulletin103108.pdf.

71 Ibid.

72 Malamud to Swartz, September 30, 2008, 5:13 p.m.

73 John Schwartz, "An Effort to Upgrade a Court Archive System to Free and Easy." *New York Times*, February 12, 2009, http://www.nytimes.com/2009/02/13/us/13records.html.http://www.nytimes.com/2009/02/13/us/13records.html.

74 "Aaron Swartz's FBI File," *Firedoglake*, February 19, 2013, http://news.firedoglake.com/2013/02/19/aaron-swartzs-fbi-file/.

75 Ibid.

76 Carl Malamud to Aaron Swartz, April 14, 2009, 11:02 a.m., https://public.resource.org/aaron/pub/msg00689.html.

77 Aaron Swartz to Carl Malamud, April 14, 2009, 12:52 p.m., https://public.resource.org/aaron/pub/msg00693.html.

78 Aaron Swartz to Carl Malamud, April 15, 2009, 8:56 p.m., https://public.resource.org/aaron/pub/msg00712.html.

79 Carl Malamud to Aaron Swartz, April 15, 2009, 9:01 p.m., https://public.resource.org/aaron/pub/msg00713.html.

80 Aaron Swartz, "Wanted by the FBI," *Raw Thought*, October 5, 2009, http://www.aaronsw.com/weblog/fbifile.

81 Carl Malamud to Aaron Swartz, January 26, 2009, 6:47 p.m., https://public.resource.org/aaron/pub/msg00509.html.

82 Carl Malamud to Aaron Swartz, January 27, 2009, 11:10 a.m., https://
    public.resource.org/aaron/pub/msg00511.html.
83 Carl Malamud to Aaron Swartz, January 27, 2009, 11:55 a.m., https://
    public.resource.org/aaron/pub/msg00513.html.
84 Ibid.
85 "Projects," *Content Liberation Front*, crawled on August 29, 2009, https://
    web.archive.org/web/20090829123824/http://contentliberation.com/.

8. HACKS AND HACKERS

1 Schonfeld, *JSTOR*, 3.
2 Ibid., 13.
3 Ibid., 131.
4 Ibid., 33.
5 Ibid., 229.
6 *JSTOR Evidence in United States vs. Aaron Swartz*, http://docs.jstor.org
   /zips/JSTOR-Swartz-Evidence-All-Docs.pdf, 19 (hereafter referred to as
   JSTOR Documents).
7 Ibid., 130.
8 Ibid., 25.
9 Ibid., 26.
10 Ibid., 7.
11 Ibid., 41.
12 Ibid., 53.
13 Ibid., 45.
14 Ibid., 130.
15 Ibid., 118.
16 Ibid., 134.
17 Ibid., 77.
18 Ibid., 97.
19 Ibid., 130.
20 Ibid., 138.
21 Ibid., 134.
22 Ibid., 167.
23 Ibid., 173.
24 Ibid., 170.
25 Ibid., 220.
26 Ibid., 236.

27 Ibid., 365–66.

28 Ibid., 320–21.

29 Ibid., 197.

30 Ibid., 228.

31 Ibid., 174.

32 Ibid., 379.

33 Internet at Liberty conference, https://www.events-google.com/google /frontend/reg/tOtherPage.csp?pageID=18711&eventID=79&event ID=79.

34 Noam Scheiber, "The Inside Story of Why Aaron Swartz Broke into MIT and JSTOR," *New Republic*, February 13, 2013, http://www.newrepublic .com/article/112418/aaron-swartz-suicide-why-he-broke-jstor-and-mit.

35 Wikler, interview.

36 Aaron Swartz, "Last Goodbyes," *Raw Thought*, June 19, 2008, http://www .aaronsw.com/weblog/lastgoodbyes.

37 Aaron Swartz, "Moving On," *Raw Thought*, June 16, 2008, http://www .aaronsw.com/weblog/movingon.

38 Aaron Swartz, "A Political Startup," *Raw Thought*, September 8, 2009, http://www.aaronsw.com/weblog/pcccstory.

39 Ibid.

40 Swartz, "SFP: Come see us."

41 Abelson et al., *MIT Report*, 30.

42 Aaron Swartz, "A Life Offline," *Raw Thought*, May 18, 2009, http://www .aaronsw.com/weblog/offline.

43 Aaron Swartz, "My Life Offline," *Raw Thought*, July 24, 2009, http://www .aaronsw.com/weblog/offline2.

44 "Aaron Swartz," Edmond J. Safra Center for Ethics, http://ethics.harvard .edu/people/aaron-swartz.

45 Aaron Swartz, "Life in a World of Pervasive Immorality: The Ethics of Being Alive," *Raw Thought*, August 2, 2009, http://www.aaronsw.com /weblog/immoral.

46 Aaron Swartz, "A Short Course in Ethics," *Raw Thought*, September 14, 2009, http://www.aaronsw.com/weblog/ethics.

47 Aaron Swartz, "Honest Theft," *Raw Thought*, September 15, 2009, http:// www.aaronsw.com/weblog/honesttheft.

48 Joss Winn, "Hacking in the University: Contesting the Valorisation of Academic Labour," *tripleC* 11, no. 2 (2013): 487.

49 Massachusetts Institute of Technology, *President's Report*, 12.

50 Ibid., 14.

51 S. S. Schweber, "Big Science in Context: Cornell and MIT," in *Big Science*, eds. Galison and Hevly, 151.

52 Ibid., 156.

53 MIT Department of Physics, "Physics and War: 1940–1945," accessed on July 11, 2015, http://web.mit.edu/physics/about/history/1940-1945.html. Also see Kevles, *Physicists*.

54 Schweber, "Big Science in Context," 152.

55 Bush, *Science—The Endless Frontier*, 56.

56 Leslie, *Cold War*, 9.

57 Merton, *Sociology of Science*, 270.

58 Ibid., 273.

59 Ibid., 275.

60 "MIT and Industry," Massachusetts Institute of Technology, accessed August 20, 2015, http://web.mit.edu/facts/industry.html.

61 Hatakenaka, "Flux and Flexibility," 100.

62 Ibid., 59.

63 Ibid., 131.

64 Washburn, *University, Inc.*, 102.

65 Merton, *The Sociology of Science*, 272.

66 "MIT Media Lab and Bank of America Announce Center for Future Banking," *MIT News*, March 31, 2008, http://newsoffice.mit.edu/2008/banking-0331.

67 Aaron Swartz, "Boston Trip Story," *Aaron Swartz: The Weblog*, April 25, 2002, http://www.aaronsw.com/weblog/000233.

68 JSTOR Documents, 1301.

69 Ibid., 1302.

70 Aaron Swartz, "r | p 2010: The Social Responsibility of Computer Science," YouTube, speech delivered on October 16, 2010, https://www.youtube.com/watch?v=SDetZaIKmPI.

71 Stinebrickner-Kauffman, interview.

72 Wikler, interview.

73 JSTOR Documents, 1299.

74 Ibid., 3089.

75 Aaron Swartz FOIA Release, Secret Service Documents, 880–81 (hereafter referred to as Swartz FOIA).

76 Ibid., 594, 639, 883.

77 JSTOR Documents, 3093.

78 Swartz FOIA, 2526.

79 Ibid., 2505.

80 Ibid., 2507.

81 Ibid., 703.

82 Ibid., 2492.

83 Abelson et al., *MIT Report*, 24.

84 Ibid., 25.

85 Swartz FOIA, 2634.

86 Aaron Swartz, Twitter post, January 6, 2011, 8:24 a.m., https://twitter.com /aaronsw.

87 JSTOR Documents, 3105.

88 Ibid., 3107.

89 Stephen Heymann, "Legislating Computer Crime," in "Computer Legislation," special issue, *Harvard Journal on Legislation* 34 (1997): 382.

90 Swartz, "SFP: Come see us."

91 JSTOR Documents, 3249–50.

9. THE WEB IS YOURS

1 A version of this paragraph originally appeared in *Slate* and is used here, gratefully, with permission.

2 Aaron Swartz, "The 2011 Review of Books," *Raw Thought*, April 19, 2012, http://www.aaronsw.com/weblog/books2011.

3 Norton, "Life Inside the Aaron Swartz Investigation."

4 Swartz FOIA, 528–92.

5 Ibid., 1590.

6 Carl Malamud to Aaron Swartz, February 6, 2011, 6:38 p.m., https:// public.resource.org/aaron/pub/msg00825.html.

7 Swartz FOIA, 731.

8 Ibid., 728.

9 Ibid., 2469.

10 Aaron Swartz to Stephen Schultze, February 1, 2011, 3:23 p.m., https:// public.resource.org/aaron/pub/msg00818.html.

11 Swartz FOIA, 1871.

12 Ibid., 458.

13 Ibid., 97.

14 Abelson et al., *MIT Report*, 31.

15 "Aaron Swartz—How Congress Works," Edmond J. Safra Center for Ethics, http://ethics.harvard.edu/aaron-swartz-%C2%A0%E2%80%94%C2%A0 -how-congress-works.

16 "Senators Introduce Bipartisan Bill to Combat Online Infringement," press release, September 20, 2010, http://www.leahy.senate.gov/press /senators-introduce-bipartisan-bill-to-combat-online-infringement.

17 Aaron Swartz, interview by Ben Wikler, *The Flaming Sword of Justice*, podcast audio, January 22, 2011, http://www.flamingswordofjustice.com /episodes/2-2011-1968.

18 Moon, Ruffini, and Segal, *Hacking Politics*, 6.

19 Swartz FOIA, 667.

20 Norton, "Life Inside the Aaron Swartz Investigation."

21 Swartz FOIA, 25.

22 Norton, "Life Inside the Aaron Swartz Investigation."

23 Swartz FOIA, 790–95; also 856–62.

24 Norton, "Life Inside the Aaron Swartz Investigation."

25 Ibid.

26 Ryan Singel, "Aaron Swartz and the Two Faces of Power," *Wired*, January 18, 2013, http://www.wired.com/2013/01/aaron-swartz-activism-and -the-two-sided-sword-of-power/.

27 Swartz FOIA, 863.

28 Ibid., 796–99; also 868–72.

29 "COICA twice defeated has offspring bent on revenge," *Demand Progress Blog*, May 11, 2011, https://web.archive.org/web/20120609221325 /http://blog.demandprogress.org/2011/05/coica-twice-defeated-has -offspring-bent-on-revenge/.

30 Senator Patrick Leahy, "Senate Judiciary Committee Unanimously Approves Bipartisan Bill to Crack Down on Rogue Websites," news release, May 26, 2011, http://www.leahy.senate.gov/press/senate-judiciary -committee-unanimously-approves-bipartisan-bill-to-crack-down-on -rogue-websites.

31 "Wyden Places Hold on Protect IP Act," news release, May 26, 2011, http://www.wyden.senate.gov/news/press-releases/wyden-places-hold-on -protect-ip-act.

32 Moon, Ruffini, and Segal, *Hacking Politics*, 81.

33 Swartz FOIA, 866.

34 Abelson et al., *MIT Report*, 42.

35 Ibid.

36 Ibid., 58.

37 Ibid., 59–60.

38 Ibid., 52–54.

39 Ibid., 59.

40 Alexis C. Madrigal, "Editor's Note to Quinn Norton's Account of the Aaron Swartz Investigation," *Atlantic*, March 3, 2013, http://www.the atlantic.com/technology/archive/2013/03/editors-note-to-quinn-nortons -account-of-the-aaron-swartz-investigation/273666/.

41 Norton, "Life Inside the Aaron Swartz Investigation."

42 Taren Stinebrickner-Kauffman, "Who Am I?," *TarenSK*, January 31, 2013, http://tarensk.tumblr.com/post/41956434294/who-am-i.

43 Stinebrickner-Kauffman, interview.

44 Ibid.

45 Ibid.

46 Holden Karnofsky, "In Memory of Aaron Swartz," *The GiveWell Blog*, January 16, 2013, http://blog.givewell.org/2013/01/16/in-memory-of-aaron -swartz/.

47 Taren Stinebrickner-Kauffman, "MIT Memorial Service," *TarenSK*, March 13, 2013, http://tarensk.tumblr.com/post/45281114505/mit-memorial -service.

48 Ibid.

49 See Snejana Farberov, Helen Pow, and James Nye, "Surveillance Shot that ruined tragic Reddit co-founder Aaron Swartz's life," *Daily Mail*, January 13, 2013, http://www.dailymail.co.uk/news/article-2261840 /Aaron-Swartz-MIT-surveillance-shot-ruined-tragic-Reddit-founders -life.html.

50 "Federal Government Indicts Former Demand Progress Executive Director for Downloading Too Many Journal Articles," *Demand Progress Blog*, July 19, 2011, https://web.archive.org/web/20110721132939/http:// blog.demandprogress.org/2011/07/federal-government-indicts-former -demand-progress-executive-director-for-downloading-too-many-journal -articles/.

51 "Show Your Support for Aaron," *Demand Progress*, July 22, 2011, https:// web.archive.org/web/20110722104427/http://act.demandprogress.org /sign/support_aaron.

52 "UPDATE: More Than 15,000 People Sign Petition in Support of Aaron Swartz," *Demand Progress Blog*, July 19, 2011, https://web.archive.org /web/20110723205147/http://blog.demandprogress.org/2011/07/update -more-than-15000-people-sign-petition-in-support-of-aaron-swartz/.

53 John Schwartz, "Open-Access Advocate Is Arrested for Huge Download," *New York Times*, July 19, 2011, http://www.nytimes.com/2011/07/20/us /20compute.html.

54 Milton J. Valencia, "Activist Charged with Hacking," *Boston Globe*, July 20, 2011, http://www.boston.com/news/education/higher/articles/2011/07 /20/activist_charged_with_hacking_mit_network_to_download_files/.

55 "More Than 35,000 Sign Petition in Support of Aaron Swartz," *Demand Progress Blog*, July 20, 2011, https://web.archive.org/web/20110723204917 /http://blog.demandprogress.org/2011/07/more-than-35000-sign-petition -in-support-of-aaron-swartz/.

56 Abelson et al., *MIT Report*, 68.

57 Brewster Kahle, "Michael Hart of Project Gutenberg Passes," *Brewster Kahle's Blog*, September 7, 2011, http://brewster.kahle.org/2011/09/07/ michael-hart-of-project-gutenberg-passes/.

58 Tim Berners-Lee, "Long Live the Web: A Call for Continued Open Standards and Neutrality," *Scientific American*, December 2010, 80–85.

59 Moon, Ruffini, and Segal, *Hacking Politics*, 103.

60 Herman, *Fight over Digital Rights*, 196.

61 "Stop Online Piracy Act, Hearing before the Committee on the Judiciary, House of Representatives, One Hundred Twelfth Congress, First Session, on H.R. 3261," serial no. 112–154, November 16, 2011, 47.

62 Ibid., 99–100.

63 Ibid., 246.

64 Ibid., 245.

65 "American Censorship Day," November 17, 2011, https://web.archive.org /web/20111117023831/http://americancensorship.org/.

66 Ibid., November 18, 2011, https://web.archive.org/web/20111118014748 /http://americancensorship.org/.

67 Moon, Ruffini, and Segal, *Hacking Politics*, 117.

68 Brewster Kahle, "12 Hours Dark: Internet Archive vs. Censorship," *Internet Archive Blogs*, January 17, 2012, https://blog.archive.org/2012/01/17 /12-hours-dark-internet-archive-vs-censorship/.

69 Senator Bob Menendez, Twitter post, January 17, 2012, 3:17 p.m., https:// twitter.com/SenatorMenendez.

70 Senator Jeff Merkley, Twitter post, January 18, 2012, 8:47 a.m., https:// twitter.com/SenJeffMerkley.

71 Senator Mark Kirk, "Kirk Announces Opposition to PROTECT IP Act," news release, January 18, 2012, http://kirk-press.enews.senate.gov/mail/ util.cfm?gpiv=2100082649.679.318&gen=1.

72 Representative Ted Poe, "I welcome [halting of SOPA] because of the bill's overreach," e-mail message, January 20, 2012, http://projects.propublica .org/sopa/P000592.html.

73 Senator Patrick Leahy, "Comment of Senator Patrick Leahy on Postpone-
    ment of the Vote on Cloture on the Motion to Proceed to the PROTECT
    IP Act," news release, January 20, 2012, http://www.leahy.senate.gov/press
    /comment-of-senator-patrick-leahy-on-postponement-of-the-vote-on
    -cloture-on-the-motion-to-proceed-to-the-protect-ip-act.
74 Swartz, *Flaming Sword of Justice*.
75 Interview with Holmes Wilson, January 2013.
76 Swartz, *Flaming Sword of Justice*.
77 Stinebrickner-Kauffman, interview.
78 Aaron Swartz, "How We Stopped SOPA" (keynote speech, F2C: Freedom
    to Connect, Washington, DC, May 21, 2012).

10. HOW TO SAVE THE WORLD

1   Swartz, "How to Save the World, Part 1."
2   Wikler, interview.
3   Ibid.
4   Aaron Swartz, Twitter post, June 1, 2012, 10:54 a.m., https://twitter.com
    /aaronsw.
5   Aaron Swartz, "Perfect Institutions," *Raw Thought*, June 8, 2012, http://
    www.aaronsw.com/weblog/perfectinstitutions.
6   Stinebrickner-Kauffman, interview.
7   Abelson et al., *MIT Report*, 65.
8   Ibid., 66.
9   Aaron Swartz, "Do I have too much faith in science?" *Raw Thought*,
    August 10, 2012, http://www.aaronsw.com/weblog/sciencefaith.
10 Stinebrickner-Kauffman, interview.
11 Aaron Swartz, "Take a step back," *Raw Thought*, August 18, 2012, http://
    www.aaronsw.com/weblog/stepback.
12 Aaron Swartz, "A cure for nearsightedness?" *Aaron Swartz: The Weblog*,
    July 20, 2002, http://www.aaronsw.com/weblog/000432.
13 Aaron Swartz, "Believe you can change," *Raw Thought*, August 18, 2012,
    http://www.aaronsw.com/weblog/dweck.
14 Swartz, "Take a step back."
15 Aaron Swartz, "Lean into the pain," *Raw Thought*, September 1, 2012,
    http://www.aaronsw.com/weblog/dalio.
16 Interview with Jeffrey Mayersohn, January 2013.
17 Swartz, "Lean into the pain."

18 Seth Finkelstein, "Aaron Swartz 'JSTOR' case indictment revised/expanded," *Infothought*, September 14, 2012, http://sethf.com/infothought/blog /archives/001476.html.

19 US Department of Justice, *United States Attorneys' Manual 9–27.320(B)* (1997), available online at http://www.justice.gov/usam/united-states -attorneys-manual.

20 Aaron Swartz, "Since power over human beings is shown . . . ," *Raw Meat*, September 13, 2012, http://qblog.aaronsw.com/post/31460484751/since -power-over-human-beings-is-shown-in-making.

21 Abelson et al., *MIT Report*, 69.

22 Aaron Swartz, "What Happens in *The Dark Knight*," *Raw Thought*, November 1, 2012, http://www.aaronsw.com/weblog/tdk.

23 Ibid.

24 Alex Stamos, "The Truth about Aaron Swartz's 'Crime,'" *Unhandled Exception*, January 12, 2013, http://unhandled.com/2013/01/12/the-truth -about-aaron-swartzs-crime/.

25 Swartz FOIA, 738–48.

26 Interview with Elliot Peters, February 2013.

27 Taren Stinebrickner-Kauffman speech (Cooper Union, New York, NY, January 19, 2013).

28 Mayersohn, interview.

29 Stinebrickner-Kauffman, interview.

30 Ibid.

31 Ibid.

32 Paul McGeough, "Aaron Swartz: A Beautiful Mind," *Sydney Morning Herald*, February 1, 2014, http://www.smh.com.au/technology/technology -news/aaron-swartz-a-beautiful-mind-20140131-31hjr.html.

33 Woodhull, interview.

34 Peters, interview.

35 Stinebrickner-Kauffman, interview.

36 Wikler, interview.

37 A version of this paragraph and the next one originally appeared in *Slate*. Used with permission.

38 Peters, interview.

39 Wikler, interview.

40 Jean-Claude Guédon, "An Unforgettable Mind," Remember Aaron Swartz, January 18, 2013, https://github.com/rememberaaronsw/rememberaaronsw /blob/master/memories/_posts/2013-01-18-An-unforgettable-mind.md.

41 Brad Templeton, "What Is a Hero?" Remember Aaron Swartz, January 18, 2013, http://www.rememberaaronsw.com/memories/What-is-a-hero%3F .html.

42 Reynaldo Guimaraes, "God Bless Him," Remember Aaron Swartz, January 14, 2013, http://www.rememberaaronsw.com/memories/God-Bless -Him.html.

43 E-mails received by MIT.

44 Holmes Wilson, "Aaron Swartz Was a Friend, and We Went to His Funeral Tuesday," *projectbrainsaver*, January 18, 2013, https://projectbrainsaver .wordpress.com/2013/01/18/holmes-wilson-fight-for-the-future-aaron -swartz-was-a-friend-and-we-went-to-his-funeral-tuesday/.

45 "Declaration of Jack R. Pirozzolo in Support of the United States' Response to the Defendant's Motion to Modify Protective Order," March 29, 2013, 4.

46 Kevin Cullen, "On Humanity, a Big Failure in Aaron Swartz Case," *Boston Globe*, January 15, 2013, http://www.bostonglobe.com/metro/2013/01/15 /humanity-deficit/bj8oThPDwzgxBSHQt3tyKI/story.html.

47 Edward Tufte, "Speech at Aaron Swartz Public Memorial" (speech, Cooper Union, New York, NY, January 19, 2013).

48 Abelson et al., *MIT Report*, 93.

49 Aaron Swartz, "Fix the machine, not the person," *Raw Thought*, September 25, 2012, http://www.aaronsw.com/weblog/nummi.

50 Aaron Swartz, "Checking In," *Schoolyard Subversion*, December 23, 2001, http://web.archive.org/web/20020205111032/http://swartzfam.com /aaron/school/.

51 A version of this paragraph originally appeared in *Slate*. Used with permission.

SELECT BIBLIOGRAPHY

Abelson, Harold, Peter A. Diamond, Andrew Grosso, and Douglas W. Pfeiffer. *Report to the President: MIT and the Prosecution of Aaron Swartz.* Cambridge: Massachusetts Institute of Technology, 2013.

Anderson, Chris. *Free: The Future of a Radical Price.* New York: Hyperion, 2009.

Bellesiles, Michael A. *1877: America's Year of Living Violently.* New York: New Press, 2010.

Birrell, Augustine. *Seven Lectures on the Law and History of Copyright in Books.* New York: G. P. Putnam's Sons, 1899.

Bishop, Wallace Putnam. "The Struggle for International Copyright in the United States." PhD diss., Boston University Graduate School, 1959.

Blagden, Cyprian. *The Stationers' Company: A History, 1403–1959.* Cambridge, MA: Harvard University Press, 1960.

Boyle, James. *The Public Domain: Enclosing the Commons of the Mind.* New Haven, CT: Yale University Press, 2008.

Brand, Stewart. *The Media Lab: Inventing the Future at MIT.* New York: Viking, 1987.

Brate, Adam. *Technomanifestos: Visions from the Information Revolutionaries.* New York: Texere, 2002.

Brylawski, E. Fulton, and Abe A. Goldman, eds. and comps. *Legislative History of the 1909 Copyright Act.* 6 vols. South Hackensack, NJ: Fred B. Rothman, 1976.

Bush, Vannevar. *Science—The Endless Frontier: A Report to the President on a Program for Postwar Scientific Research.* Washington, DC: National Science Foundation, 1990.

Carnegie, Andrew. *Autobiography of Andrew Carnegie*. Boston: Houghton Mifflin, 1920.

Casper, Scott E., Jeffrey D. Groves, Stephen W. Nissenbaum, and Michael Winship, eds. *A History of the Book in America, Volume 3: The Industrial Book, 1840–1880*. Chapel Hill: Published in association with the American Antiquarian Society by the University of North Carolina Press, 2007.

Charvat, William. *The Profession of Authorship in America, 1800–1870*. Edited by Matthew J. Bruccoli. New York: Columbia University Press, 1992.

———. *Literary Publishing in America, 1790–1850*. Amherst: The University of Massachusetts Press, 1993.

Chomsky, Noam. *Understanding Power: The Indispensable Chomsky*. New York: New Press, 2002.

City of Boston Statistics Department. *Boston Statistics, 1922, with Memorable Sites and Buildings, Etc.* Boston: City Printing Department, 1922.

Clark, Aubert J. *The Movement for International Copyright in Nineteenth Century America*. Washington, DC: Catholic University of America Press, 1960.

Cobbett, William. *Porcupine's Works; Containing Various Writings and Selections, Exhibiting a Faithful Picture of the United States of America . . .* London: Cobbett and Morgan, 1801.

Cole, John Y. *For Congress and the Nation: A Chronological History of the Library of Congress*. Washington, DC: Library of Congress, 1979.

Coll, Gary R. "Noah Webster: Journalist, 1783–1803." PhD diss., Southern Illinois University, 1971.

Cope, Bill, and Angus Phillips, eds. *The Future of the Academic Journal*. Oxford, UK: Chandos Publishing, 2009.

Dalzell, Robert F., Jr., and Lee Baldwin Dalzell. *George Washington's Mount Vernon: At Home in Revolutionary America*. New York: Oxford University Press, 1998.

Daniel, Marcus Leonard. "'Ribaldry and Billingsgate': Popular Journalism, Political Culture and the Public Sphere in the Early Republic." PhD diss., Princeton University, 1998.

Davis, Donald G., Jr., Kenneth E. Carpenter, Wayne A. Wiegand, and Jane Aikin. *Winsor, Dewey, and Putnam: The Boston Experience*. Champaign: Graduate School of Library and Information Science, University of Illinois at Urbana-Champaign, 2002.

Dickens, Charles. *The Letters of Charles Dickens*. Edited by Mamie Dickens and Georgina Hogarth. 2 vols. New York: Charles Scribner's Sons, 1879.

Eisenstein, Elizabeth L. *The Printing Revolution in Early Modern Europe*. 2nd ed. New York: Cambridge University Press, 2005.

Eremo: Historical Memories of the Hermitage of the Massaccio's Caves. Cupramontana, Italy: Eremo srl, 2007. Based on *Le grotte dei frati bianchi* by B. Tesei.

Fleming, Edward McClung. *R. R. Bowker: Militant Liberal*. Norman: University of Oklahoma Press, 1952.

Fletcher, William Isaac. *Public Libraries in America*. Boston: Roberts Brothers, 1894.

Foner, Philip S. *The Great Labor Uprising of 1877*. New York: Monad Press, 1977.

Ford, Emily Ellsworth Fowler, comp. *Notes on the Life of Noah Webster*. 2 vols. Edited by Emily Ellsworth Ford Skeel. New York: privately printed, 1912.

Forster, John. *The Life of Charles Dickens*. 3 vols. Boston: James R. Osgood, 1875.

Galison, Peter, and Bruce Hevly, eds. *Big Science: The Growth of Large-Scale Research*. Stanford, CA: Stanford University Press, 1992.

Garrison, Lora Doris. "Cultural Missionaries: A Study of American Public Library Leaders, 1876–1910." PhD diss., University of California, Irvine, 1973.

Gilder, Richard Watson. *Letters of Richard Watson Gilder*. Edited by Rosamond Gilder. Boston: Houghton Mifflin, 1916.

Gillies, James, and Robert Cailliau. *How the Web Was Born*. New York: Oxford University Press, 2000.

Golway, Terry. *Machine Made: Tammany Hall and the Creation of Modern American Politics*. New York: Liveright, 2014.

Greenspan, Ezra. *George Palmer Putnam: Representative American Publisher*. University Park: Pennsylvania State University Press, 2000.

Gross, Robert A., and Mary Kelley, eds. *A History of the Book in America, Volume 2: An Extensive Republic: Print, Culture, and Society in the New Nation, 1790–1840*. Chapel Hill: Published in association with the American Antiquarian Society by the University of North Carolina Press, 2010.

Grossman, George S., comp. *Omnibus Copyright Revision Legislative History*. 17 vols. Buffalo, NY: William S. Hein, 1976.

Hackenberg, Michael, ed. *Getting the Books Out: Papers of the Chicago Conference on the Book in 19th-Century America*. Washington, DC: Center for the Book, Library of Congress, 1987.

Hafner, Katie, and Matthew Lyon. *Where Wizards Stay Up Late: The Origins of the Internet*. New York: Simon and Schuster, 1996.

Hannay, David. *Life of Frederick Marryat*. London: Walter Scott, 1889.

Harper, J. Henry. *The House of Harper: A Century of Publishing in Franklin Square*. New York: Harper and Brothers, 1912.

Hatakenaka, Sachi. "Flux and Flexibility: A Comparative Institutional Analysis of Evolving University-Industry Relationships in MIT, Cambridge, and Tokyo." PhD diss., Massachusetts Institute of Technology, 2002.

Herman, Bill D. *The Fight over Digital Rights: The Politics of Copyright and Technology.* New York: Cambridge University Press, 2013.

Hofstadter, Richard. *The Age of Reform.* New York: Knopf, 1955.

Holt, Henry. *Garrulities of an Octogenarian Editor.* Boston: Houghton Mifflin, 1923.

Imfeld, Cassandra Jacqueline. "Repeated Resistance to New Technologies: A Case Study of the Recording Industry's Tactics to Protect Copyrighted Works in Cyberspace between 1993 and 2003." PhD diss., University of North Carolina at Chapel Hill, 2004.

Jackall, Robert. *Moral Mazes: The World of Corporate Managers.* New York: Oxford University Press, 1988.

Johnson, Ann K. "The People's Book: Making, Selling, and Reading Reference Works in Nineteenth-Century America." PhD diss., University of Southern California, 2014.

Johnson, Edgar. *Charles Dickens: His Tragedy and Triumph.* 2 vols. New York: Little, Brown, 1952.

Johnson, Robert Underwood. *Remembered Yesterdays.* Boston: Little, Brown, 1923.

Jones, Elisabeth A. "Constructing the Universal Library." PhD diss., University of Washington, 2014.

Kaestle, Carl F., and Janice A. Radway, eds. *A History of the Book in America, Volume 4: Print in Motion: The Expansion of Publishing and Reading in the United States, 1880–1940.* Chapel Hill: Published in association with the American Antiquarian Society by the University of North Carolina Press, 2009.

Kaplan, Benjamin. *An Unhurried View of Copyright.* New York: Columbia University Press, 1967.

Kendall, Joshua. *The Forgotten Founding Father: Noah Webster's Obsession and the Creation of an American Culture.* New York: Putnam, 2010.

Kevles, Daniel J. *The Physicists: The History of a Scientific Community in Modern America.* New York: Knopf, 1977.

Knowlton, John D., ed. *Herbert Putnam: A 1903 Trip to Europe.* Lanham, MD: Scarecrow Press, 2005.

Koch, Theodore Wesley. *A Book of Carnegie Libraries.* White Plains, NY: H. W. Wilson, 1917.

———. *War Service of the American Library Association.* 3rd ed. Washington, DC: ALA War Service, 1918.

Kochen, Manfred, ed. *The Growth of Knowledge: Readings on Organization and Retrieval of Information.* New York: John Wiley and Sons, 1967.

Lee, Alfred M. *The Daily Newspaper in America.* New York: Macmillan, 1937.

Leslie, Stuart W. *The Cold War and American Science: The Military-Industrial Complex at MIT and Stanford.* New York: Columbia University Press, 1993.

Lessig, Lawrence. *Free Culture: How Big Media Uses Technology and the Law to Lock Down Culture and Control Creativity.* New York: Penguin, 2004.

Levy, Steven. *Hackers: Heroes of the Computer Revolution.* New York: Penguin, 2001. First published in 1984 by Doubleday.

Lewis, Lawrence. *A Tribute to Dr. Herbert Putnam, Librarian of Congress.* Washington, DC: US Government Printing Office, 1939.

Library of Congress. *Herbert Putnam, 1861–1955: A Memorial Tribute.* Washington, DC: Library of Congress, 1956.

Licklider, J. C. R. *Libraries of the Future.* Cambridge, MA: MIT Press, 1965.

Ludlow, Peter, ed. *High Noon on the Electronic Frontier: Conceptual Issues in Cyberspace.* Cambridge, MA: MIT Press, 1996.

Malamud, Carl. *Exploring the Internet: A Technical Travelogue.* Englewood Cliffs, NJ: Prentice-Hall, 1993.

———. *A World's Fair for the Global Village.* Cambridge, MA: MIT Press, 1997.

Markoff, John. *What the Dormouse Said: How the Sixties Counterculture Shaped the Personal Computer Industry.* New York: Viking Penguin, 2005.

Marryat, Florence. *Life and Letters of Captain Marryat.* 2 vols. New York: D. Appleton, 1872.

Marryat, Frederick. *Second Series of a Diary in America, with Remarks on Its Institutions.* Philadelphia: T. K. and P. G. Collins, 1840.

———. *Diary in America.* Edited by Jules Zanger. Bloomington: Indiana University Press, 1960.

Massachusetts Institute of Technology. *President's Report: January 1920.* Cambridge, MA: Technology Press, 1920.

Mathews, Cornelius. *The Various Writings of Cornelius Mathews.* New York: Harper and Brothers, 1843.

McGrane, Reginald Charles. *The Panic of 1837.* Chicago: University of Chicago Press, 1924.

McLeod, Kembrew. *Pranksters: Making Mischief in the Modern World.* New York: NYU Press, 2014.

McSherry, Corynne. *Who Owns Academic Work? Battling for Control of Intellectual Property.* Cambridge, MA: Harvard University Press, 2001.

Merton, Robert K. *The Sociology of Science: Theoretical and Empirical Investigations*. Edited by Norman W. Storer. Chicago: University of Chicago Press, 1973.

Micklethwait, David. *Noah Webster and the American Dictionary*. Jefferson, NC: McFarland, 2000.

Miller, Perry. *The Raven and the Whale: The War of Words and Wits in the Era of Poe and Melville*. New York: Harcourt, Brace, 1956.

Monaghan, E. Jennifer. *A Common Heritage: Noah Webster's Blue-Back Speller*. Hamden, CT: Archon Books, 1983.

———. *Learning to Read and Write in Colonial America*. Amherst: University of Massachusetts Press, 2005.

Moon, David, Patrick Ruffini, and David Segal, eds. *Hacking Politics: How Geeks, Progressives, the Tea Party, Gamers, Anarchists and Suits Teamed Up to Defeat SOPA and Save the Internet*. New York: OR Books, 2013.

Nichols, Frederick, comp. *1883 Triennial, June 1886, Secretary's Report No. 2*. Cambridge, MA: privately printed for the Harvard University Class of 1883, 1886.

Okerson, Ann Shumelda, and James J. O'Donnell, eds. *Scholarly Journals at the Crossroads: A Subversive Proposal for Electronic Publishing*. Washington, DC: Office of Scientific & Academic Publishing, Association of Research Libraries, 1995.

Overhage, Carl F. J., and R. Joyce Harman, eds. *INTREX: Report of a Planning Conference on Information Transfer Experiments*. Cambridge, MA: MIT Press, 1965.

Patry, William. *Moral Panics and the Copyright Wars*. New York: Oxford University Press, 2009.

Peek, Robin P., and Gregory Newby, eds. *Scholarly Publishing*. Cambridge, MA: MIT Press, 1996.

Putnam, George Haven, ed. *The Question of Copyright*. New York: G. P. Putnam's Sons, 1891.

———. *George Palmer Putnam: A Memoir*. New York: G. P. Putnam's Sons, 1912.

———. *Memories of a Publisher, 1865–1915*. New York: G. P. Putnam's Sons, 1915.

Putnam, George Palmer. *The tourist in Europe; or, A concise summary of the various routes, objects of interest &c in Great Britain, France, Switzerland . . .* New York: Wiley and Putnam, 1838.

Register of Debates in Congress, Comprising the Leading Debates and Incidents of the Second Session of the Twenty-Fourth Congress . . . Vol. 13. Washington, DC: Gales and Seaton, 1837.

Report of the Trustees of the Public Library of the City of Boston. Boston: J. H. Eastburn, City Printer, 1852.

Reports of Committees of the House of Representatives for the First Session of the Fifty-First Congress, 1889–'90. 10 vols. Washington, DC: Government Printing Office, 1891.

Rollins, Richard M. *The Long Journey of Noah Webster.* Philadelphia: University of Pennsylvania Press, 1980.

Rose, Mark. *Authors and Owners: The Invention of Copyright.* Cambridge, MA: Harvard University Press, 1993.

Rosenberg, Jane Aikin. *The Nation's Great Library: Herbert Putnam and the Library of Congress, 1899–1939.* Urbana: University of Illinois Press, 1993.

Schonfeld, Roger C. *JSTOR: A History.* Princeton, NJ: Princeton University Press, 2003.

Scudder, Horace E. *Noah Webster.* Boston: Houghton Mifflin, 1889.

Shove, Raymond H. "Cheap Book Production in the United States, 1870 to 1891." MA diss., University of Illinois, 1937.

Smith, Helen Evertson. *Colonial Days & Ways.* New York: Century, 1900.

Snyder, K. Alan. *Defining Noah Webster: Mind and Morals in the Early Republic.* Lanham, MD: University Press of America, 1990.

Solberg, Thorvald. *International Copyright in the Congress of the United States, 1837–1886.* Boston: Rockwell and Churchill, 1886.

———. *Copyright in Congress, 1789–1904.* Washington, DC: Government Printing Office, 1905.

Stedman, Laura, and George M. Gould, MD, eds. *Life and Letters of Edmund Clarence Stedman.* 2 vols. New York: Moffat, Yard, 1910.

Stefik, Mark, ed. *Internet Dreams: Archetypes, Myths, and Metaphors.* Cambridge, MA: MIT Press, 1996.

Stein, Allen F. *Cornelius Mathews.* New York: Twayne Publishers, 1974.

Stern, Madeleine B., ed. *Publishers for Mass Entertainment in Nineteenth Century America.* Boston: G. K. Hall, 1980.

"Stop Online Piracy Act, Hearing before the Committee on the Judiciary, House of Representatives, One Hundred Twelfth Congress, First Session, on H.R. 3261." Serial no. 112-154, November 16, 2011.

Tirella, Joseph. *Tomorrow-Land: The 1964–65 World's Fair and the Transformation of America.* Guilford, CT: Lyons Press, 2014.

Townley, Benjamin. *The Quest for Nationality: An American Literary Campaign.* Syracuse, NY: Syracuse University Press, 1957.

Turner, Fred. *From Counterculture to Cyberculture: Stewart Brand, the Whole Earth Network, and the Rise of Digital Utopianism*. Chicago: University of Chicago Press, 2006.

Twain, Mark. *Autobiography of Mark Twain*. Edited by Benjamin Griffin and Harriet Elinor Smith. 2 vols. Berkeley: University of California Press, 2013.

Unger, Harlow Giles. *Noah Webster: The Life and Times of an American Patriot*. New York: John Wiley and Sons, 1998.

US Senate Committee on the Judiciary. *The Copyright Term Extension Act of 1995 Hearing, 20 September 1995*. Washington, DC: Government Printing Office, 1997.

Vaidhyanathan, Siva. *The Anarchist in the Library: How the Clash Between Freedom and Control Is Hacking the Real World and Crashing the System*. New York: Basic, 2004.

Van Dyke, Henry. *The National Sin of Literary Piracy: A Sermon Preached by Henry Van Dyke*. New York: C. Scribner's Sons, 1888.

Van Slyck, Abigail Ayres. "Free to All: Carnegie Libraries and the Transformation of American Culture, 1886–1917." PhD diss., University of California, Berkeley, 1989.

Waldrop, M. Mitchell. *The Dream Machine: J. C. R. Licklider and the Revolution That Made Computing Personal*. New York: Viking, 2001.

Warfel, Harry R. *Noah Webster: Schoolmaster to America*. New York: Macmillan, 1936.

Washburn, Jennifer. *University, Inc.: The Corporate Corruption of Higher Education*. New York: Perseus, 2005.

Webster, Noah. *A Collection of Essays and Fugitiv [sic] Writings on Moral, Historical, Political, and Literary Subjects*. Boston: I. Thomas and E. T. Andrews, 1790.

———. *A Collection of Papers on Political, Literary, and Moral Subjects*. New York: Webster and Clark, 1843.

———. *Poems by Noah Webster*. Edited by Ruth Farquhar Warfel and Harry Redcay Warfel. College Park, MD: Harruth Lefraw, 1936.

———. *Letters of Noah Webster*. Edited by Harry R. Warfel. New York: Library Publishers, 1953.

———. *The Autobiographies of Noah Webster*. Edited by Richard M. Rollins. Columbia: University of South Carolina, 1989.

Weeks, Lyman Horace. *A History of Paper-Manufacturing in the United States, 1690–1916*. New York: Lockwood Trade Journal Company, 1916.

Williams, Julie Kay Hedgepeth. "The Significance of the Printed Word in Early America: Colonists' Thoughts on the Role of the Press." PhD diss., University of Alabama, 1997.

Williams, Sam. *Free as in Freedom (2.0): Richard Stallman and the Free Software Revolution*. Boston: Free Software Foundation, 2010.

Wilson, Christopher P. *The Labor of Words: Literary Professionalism in the Progressive Era*. Athens: University of Georgia Press, 1985.

Wirtén, Eva Hemmungs. *No Trespassing: Authorship, Intellectual Property Rights, and the Boundaries of Globalization*. Toronto: University of Toronto Press, 2004.

INDEX

academic research:
  and copyright, 11, 88–89
  corporate funding of, 209–12, 265
  and gift economy vs. market
    economy, 89
  government funding of, 174,
    208–9, 211
  and open access, 176–77
  peer review of, 174–75
  publishers of, 175–77, 178
  results harvested and sold, 212
  serials pricing crisis, 175
  Swartz's robotic harvesting of,
    197–202
  and technology transfer offices, 212
Adams, Henry, 61
Adams, John, 32
Adida, Ben, 130
Advanced Research Projects Agency
  (ARPA), 82, 100–101
Æolian Company, 75–76
AI (artificial intelligence), 102–4,
  112, 125
Aikin, Jane, 71

Alcott, Louisa May, 55, 60, 117–18
Alden, John Berry, 58
Alexa Internet, 135
Amazon, 241
American Bar Association, 86
American colonies:
  cultural independence for, 23
  freedom of expression in, 20
  nationalism in, 20
  national language for, 22–23
  printed word valued in, 27
  printing trade in, 20
American Copyright League, 60, 64
*American Dictionary of the English
  Language* (Webster), 34, 35–37
American Graphophone Company,
  74–75
American Library Association:
  and e-books, 107
  Library/USA, 80–81, 82
  Library War Service, 77
American Minerva, The (newspaper),
  32
American News Company, 55

American Publishers' Copyright
    League, 60–61, 62
American Revolution, 20, 22, 25
American Society of Composers,
    Authors and Publishers
    (ASCAP), 84
Andrew W. Mellon Foundation,
    195–96
Anonymous (hacking collective), 263
Apple:
    iTunes, 153, 178
    Macintosh, 106, 147
    Swartz's essay on, 267–68
Appleton, William, 63
Arab Spring, 237
archive.org, 193
ARPANET, 101–2
ArsDigita Prize, 124
art, copyright protections of, 73
Articles of Confederation, 26
arXiv, 176–77
ASCII, 109
Ashcroft, John, 122–23, 139
AT&T, 264
Atlantic Monthly, 65, 82, 228
authors, *see* creators
Authors League of America, 84
Avaaz, 203, 241, 248, 249

Baffler (magazine), 6, 229
Banff Forum, 166
Bank of America, 212–13
Barday, Shireen, 173
Barlow, Joel, 26, 33
Bayh-Dole Act (1980), 211–12
Beadle Co., 52
Beged-Dov, Gabe, 126
Berkman Center for Internet &
    Society, 121

Berlin Declaration, 177
Berners-Lee, Tim, 3, 9, 107–8, 127,
    128, 143, 237, 238
Bethedsa Statement, 177
Bill of Rights, 105
Birrell, Augustine, 18–19
BitTorrent, 131
Blake, William, 99
Blue Book, 181–82
"blue box," 263–64
Boing Boing (blog), 131, 139
Bono, Mary, 118, 138
Book People, 120–21, 138, 140
books:
    archiving of, 135–36
    e-books, 99, 107, 117
    sharing our cultural legacy, 164
Boston Public Library, 67, 139
Bowen, William G., 195–96
Bracha, Oren, 30
Bradbury, Ray, 80
    Fahrenheit 451, 120–21
Brand, Stewart, *The Media Lab*, 12,
    269
Brown, Ralph S. Jr., 85–86
Buchanan, James, 47
Buchheit, Paul, 165
Budapest Declaration, 177
Buffett, Warren, 7
Burke, Colin, 91
Bush, Vannevar, 82–83, 108, 208,
    209, 211

Camaldolese order, 169
Cambridge, Massachusetts, Swartz's
    residence in, 202–5, 229
Cameron, S. T., 74–75
Capra, Frank, 183
Carnegie, Andrew, 68

Carnegie Mellon University, 115
Caro, Robert, *The Power Broker*, 203
Carstensen, Simon, 149, 158
Carter, Jimmy, 211–12
Cassedy, Tim, 24
Caves of Massaccio, 169
censorship:
    copyright laws as, *see* copyright laws
    of Internet, 226, 231, 233, 238, 244
    by printers' guild, 18–19
Century Magazine, 60, 62
CERN, 107–8
Challenge to Greatness, 79–81
Change.org, 248, 249
Charles Scribner's Sons, 61
Charvat, William, 25
Cheng, Tiffiniy, 152, 240, 241
Chicago Force, 124
Chomsky, Noam, *Understanding Power*, 171–72, 181, 219
CIA, 82
City Club, Manhattan, 72
civil disobedience, 180
Clarke, Arthur C., 152
Clay, Henry, 47, 49
Clemens, Samuel (Twain), 60, 62, 73
Cleveland, Grover, 61
Clinton, Bill, 119, 241
Clinton Global Initiative, 241
Cobbett, William, 33
Cohen, Bram, 131
Cold War, 78
    end of, 211
Cole, John Y., 69
Coll, Gary R., 32

collusion, 54–55
Combating Online Infringement and Counterfeits Act (COICA, 2010), 225–27, 230–31, 237
Communications Week, 182
communication technology, 12, 13, 18, 87–88
Computer Fraud and Abuse Act (CFAA), 218–19, 235, 237, 253
computers, 91
    Apple Macintosh, 106, 147
    code sharing, 104
    and e-books, 99, 107
    free software, 103, 104, 107, 190, 230, 266
    and hackers, 102–3, 262–63
    human-computer symbiosis, 101
    operating systems, 106
    personal, 100, 106, 147
    proprietary software, 103–4
    protected, defined, 218, 235, 253
    robotic harvesting via, 198–99
    as thinking machines, 102
    UNIVAC, 81, 90
    unsupervised misfits' use of, 98, 100, 103–5
    use of term, 96
    value added by, 97
    Xerox Sigma V mainframe, 95–97, 113
Computers and the Humanities, 96
Condé Nast, 2, 156–61, 164, 170
Congress, US:
    1790 Copyright Law, 30–31
    1831 Copyright Act, 38
    1909 Copyright Act, 76–77
    1995 Copyright Term Extension Act, 110–13, 136

Congress, US (*cont.*)
  1998 Sonny Bono Copyright
    Term Extension Act (CTEA),
    118–20, 122, 172
  and copyright law revisions, 37,
    68, 70, 72, 73–77, 269
  and corporate interests, 132
  and international copyright,
    59–65
  and NET Act, 119, 132, 237
  and PIPA, 230–31, 237–44, 247
  and public domain users, 111–12,
    132–34
  and SOPA, 238–44
Connecticut, first US copyright law
  in, 27
Connolly, Dan, 125
Constitution, US, 30, 86, 105, 136,
  138
Content Liberation Front, 193–94,
  201
Continental Congress, 26
Cooper, James Fenimore, 35
copyright:
  absolute, 129–30
  extension to new media, 71–77,
    84, 87–88
  foreign works unprotected by,
    49–50
  and gift economy vs. market
    economy, 89
  international, 43, 45–48, 49–53,
    54, 59–65, 68
  and licensing, 130–31
  as monopoly, 61, 74
  ownership of, 20, 26, 36–37, 39,
    46, 90, 179
  perpetual, 19, 26, 118–19, 138,
    139
  as property right, 5, 11, 26, 39,
    65, 223, 268
  public benefit of, 73, 76, 268
  as social relationship, 26, 45–46,
    134, 268
  as statutory right, 49, 130
  and Webster, 23–24, 27–31,
    36–40
copyright conservancy, 129
copyright laws:
  1790 Copyright Act, 4, 30–31,
    33–34
  1831 Copyright Act, 37, 43
  1909 Copyright Act, 76–77, 87
  1976 Copyright Act, 86–88,
    90–91, 110, 138
  1995 Copyright Term Extension
    Act, 110–13, 118, 132–33,
    136, 243
  1998 Sonny Bono CTEA, 118–20,
    139, 172, 269
  American, 20, 26
  as censorship, 17–19
  COICA, 225–27, 230–31, 237
  in Connecticut, 27
  debates on revisions (1965–76),
    83–88
  DMCA, 119, 132, 237
  in England, 19–20, 26, 37
  goal of, 89–90
  morality and metaphor reflected
    in, 140, 268
  NET Act, 119, 132–33, 237
  penalties for scofflaws, 13, 73
  PIPA, 230–31, 237–44, 247, 248
  and public domain, 3–4, 86, 115
  in the public interest, 46
  restrictive, 120–21, 122, 237–38,
    269

in separate states, 30
SOPA, 238–44, 248
Statute of Anne, 19–20, 30–31
works protected by, 26, 65
copyright reform movement, 124
copyright term, 26, 73, 84
   economic arguments for, 134–35
   fourteen years from publication
      date, 19–20, 30
   limits of, 136
   ninety-five years after publication,
      119
   political implications in, 172
   and public domain, 4, 110–11,
      134–35
   seventy years after author's death,
      4, 119
   twenty-eight years or life of
      author, 37–38
   twenty-eight years plus renewal,
      77, 90
Cornyn, John, 264
court documents, and PACER,
      185–93, 202, 223, 267
Craigslist, 241
Cramer, Jim, 203
Creative Commons, 129–31, 267
creators:
   copyright ownership by, 20, 26,
      36–37, 39, 46, 90
   financial risks assumed by, 25
   in gift economy vs. market
      economy, 89
   incentives for, 13, 20, 39, 40, 42,
      50, 86, 89
   non-US authors, 39, 41–47,
      49–50, 268
   rights of, 50, 85
   self-financing by, 25

US copyright movement led by,
      26–27
cultural brain, 11, 269
Cupramontana, Italy, 170, 173–74,
      177–78
cyber-crime, 5

Daniel, John W., 63
Dante Alighieri, *Divine Comedy*, 115
Dark Knight, The (movie), 255
Davis, G. Howlett, 76
Davis, Watson, 90–91
Dawes, Thomas, 35
Declaration of Independence, US,
      97, 99
Defense Department, US, 100–101
Demand Progress, 225, 226, 228,
      230–31, 234, 236, 243, 267
Dickens, Charles, 42, 49–50
Dictionary of the English Language
      (Johnson), 34
Digital Millennium Copyright Act
      (DMCA), 119, 132–33, 237
digital networks:
   closed systems in, 269
   transforming society via, 12, 122,
      266–67
Dillon, Will, 85
Dilworth, Thomas, *A New Guide to
      the English Tongue*, 22
dime novels, 52, 55
Disney Company, 118, 119
DNS blocking, 226
Doctorow, Cory, 131, 139
Document Liberation Front, 182
Dodd, Chris, 243
Dodge, A. J., 263–64
Dotcom, Kim, 226
Downhill Battle, 152–54, 155, 178

Dr. Seuss Enterprises, 134
Dunne, Finley Peter, 68
Duranceau, Ellen Finnie, 198, 215, 216
Dylan, Bob, 111

Eastman, George, 207
eBay, 241
edSITEment, 117
education, value of, 48–49
Eldred, Eric, 117–19, 120, 121, 122–23, 129, 132, 134, 138, 140
Eldred v. Ashcroft, 122–23, 124, 129, 134, 136–41, 162, 242
Eldritch Press, 117–19, 121, 237
electric lights, 69
Electronic Data Gathering and Retrieval (EDGAR), 184, 185
Electronic Frontier Foundation, 230, 261
Electronic Information for Libraries (EIFL), 173–74, 179, 190
Emerging Technologies Conference, 131–32, 136
England:
    American resentments against, 45
    class hierarchy in, 44
    copyright laws in, 19–20, 26, 37
    Glorious Revolution (1688) in, 19
    Parliament in, 19
    printers' guilds in, 18–19, 26
    Statute of Anne, 19–20, 30–31
Escher, M. C., 249

fact vs. artifact, 88–89
FBI, 191–92, 223
Federalists, 32–33
Felter, Wes, 9, 128, 131

Fight for the Future, 240, 241–42
file lockers, 226
file sharing:
    Congress suspicions of, 132
    online, 4, 152–54
    peer-to-peer, 133–34, 152–54
    as pull marketing approach, 133
    as theft, 4, 133, 137, 152–54, 179, 235–36
Finkelstein, Herman, 84
Finkelstein, Seth, 253
First Amendment, 242
Flaming Sword of Justice, The (podcast), 241, 243
Ford, Paul, 162
Forster, John, 49
4chan, 240
free culture movement, 3–4, 98, 140–41, 152–55, 167, 179, 204, 223
Freedom of Information Act (FOIA), 188, 223
freedom of speech, 20, 231, 242
Freedom to Connect conference, 244
Free Software Foundation, 104, 107, 190, 230, 266
Frost, Robert, 118, 121
Furman, Charlie, 257
Furniss, George W., 72–73

Gagarin, Yuri, 78
Gates, Bill, 106
Gay, Joshua, 230
Gilbert, Jon, 131
Gilder, Richard Watson, 60, 61–62, 64
Ginsburg, Ruth Bader, 139
Ginsparg, Paul, 176
Giustiniani, Paolo, 169–70, 174

GiveWell, 234, 249
Gladwell, Malcolm, 251
Glorious Revolution (1688), 19
GNU Project, 104, 107, 114, 154, 190
Gnutella, 133
Godey's Lady's Book, 52
Golway, Terry, *Machine Made*, 57
Good, Andrew, 229, 255
Google, 131, 185, 239
    in *Bubble City* (fiction), 164, 165
    and PIPA, 230, 241
Google Books, 173
Google Print for Libraries, 163
Gore, Al, 183
Gorton, Nathaniel M., 256–57
government:
    menace and intimidation by, 254,
        262, 264
    open, 172, 173
    public data from, 183–85
    research funded by, 82–83, 101,
        174, 208–9, 211
Graham, Paul, 145–48, 149, 219
Graham's, 52
*Grammatical Institute of the English
    Language, A* ["blue-backed
    speller"] (Webster), 23, 25, 27,
    30, 33, 34, 36
Green, James N., 25
Greenspan, Alan, 138
Greenspun, Philip, 124
Guédon, Jean-Claude, 261
Guerilla Open Access Manifesto,
    6–7, 178–81, 189–90, 201,
    228–30, 247
Guernica (online magazine), 5
Guest, Edgar A., 109
Guimaraes, Reynaldo, 262
Gutenberg, Johannes, 18, 98–99

hacker ethic, 103–4, 112, 125, 135,
    206, 212, 266–67
Hackers (Levy), 102–3, 138
Hafner, Katie, and Matthew Lyon,
    Where Wizards Stay Up Late,
    101
Hannay, David, 42
Harnad, Stevan, 176
Harper, J. Henry, 41
Harper's, 52
Harper brothers, 41–42
Harrison, Benjamin, 64
Hart, Michael, 93–100, 122, 268
    A Brief History of the Internet,
    109–10
    and copyright legislation, 120–21
    death of, 238
    documents digitized by, 97–100,
    105
    and *Eldred v. Ashcroft*, 140
    and Internet access, 96, 100
    and Project Gutenberg, 99, 105–7,
    108–9, 112–15
    resisting authority, 94–95
    thriving in nonconformist
    environments, 103, 104–5,
    267
    and Xerox Sigma V, 95–97, 113
Harvard University:
    Swartz as research affiliate at, 3,
    205, 207, 223, 229
    Swartz banned from, 224
Hatakenaka, Sachi, 210–11
Hatch, Orrin, 110, 111, 225
Hawthorne, Nathaniel, 52
    Scarlet Letter, 117
Hellman, Eric, 188
Hendler, James, 128
Henley, Don, 111

Herman, Bill D., *The Fight over
Digital Rights*, 239
Heymann, Stephen, 2, 216
   and CFAA, 218–19
   just doing his job, 264
   "Legislating Computer Crime"
      by, 218
   as prosecutor in Swartz's case,
      5, 218, 222, 229, 236–37,
      256–57, 263, 264
Hiscock, Frank, 62
Holder, Eric, 264
Hollister Co., 249
Holt, Henry, 65
Homestead labor dispute, 68
Honeycutt, Kristina, 192
Houdini, Harry, 78
Howells, William Dean, 60, 118
"How to Save the World, Part 1"
   (Swartz), 7, 247–48
Huffman, Steve, 149–50, 158–59
hypertext, 108

ideas, social value of, 17
Image Atlas, 244
Infogami, 147, 148–51, 158
Info Network, The, 124
information:
   artificial maintenance of scarcity,
      4, 177–78, 184
   and communication technology,
      12
   digitization of, 82–83
   dissemination of, 67–69, 83, 88,
      98, 115, 162, 178, 185
   fact vs. artifact, 88–89
   and public policy, 11
   sharing, 101–4, 174
   unauthorized access to, 13–14

unrelated to enlightenment, 32–34
"wants to be expensive," 13, 70,
   269
"wants to be free," 12, 70, 98,
   104, 269
"wants to be locked away," 24,
   178
information explosion, 84
information superhighway, 108, 111
Ingram, Mathew, 158
intellectual property:
   as a right, 26, 46
   and copyright laws, 26, 31, 73,
      88, 119, 226
   and fact vs. artifact, 88–89
   and gift economy vs. market
      economy, 89
   and monopoly, 46, 61
   official attitudes about, 115, 134,
      237, 268
   PIPA, 230–31
   sale or rental of, 212
   sociopolitical implications of, 24,
      268–69
   and technology, 90–91
   unauthorized access to, 13–14
International Telecommunications
   Union (ITU), 181–82
Internet:
   and academic publishing, 176
   blackouts of, 240, 241–43
   Blue Book standards manual,
      181–82
   censorship of, 226, 231, 233, 238,
      244
   changing the world via, 183, 267
   community platforms, 156, 269
   Congress suspicions of, 132
   and copyright restrictions, 119

decentralized architecture of, 100, 132

decentralized communication on, 14

early use of, 96, 100, 112

and free culture movement, 3–4, 141

history of, 109–10

inaccessible content of, 14

as infinite library, 3, 13, 163, 269

information retrieval and distribution via, 115, 125, 182, 268

infrastructure of, 108

and market economy, 120

niche medium, 182, 183

as "one big library," 91, 127–28

political implications of, 243

as profit-driven network, 174, 179

"scrapers" of, 172, 197–200

social networks on, 127, 238

terms-of-service agreements on, 218–19

usability issues, 125

and World Wide Web, 98, 108–10

Internet Archive, 135–36, 242

Internet at Liberty conference, 201

Internet Censorship Day, 240, 241

Internet's Own Boy, The (documentary), 14

Internet Wiretap, 112

Irving, Washington, 35

Ito, Joi, 232, 250

iTunes, 153, 178

Jackall, Robert, *Moral Mazes*, 157, 165

James, Henry, 117

Jaszi, Peter, 111–12

Jay, John, 35

Jefferson, Thomas, 33

Jobs, Steve, 106, 267–68

John, Richard R., 31

Johnson, Ann K., 48

Johnson, Robert Underwood, *Remembered Yesterdays*, 62–64

Johnson, Samuel, *Dictionary of the English Language*, 34

Jones, Elisabeth A., 162, 163

JSTOR, 195–202

beginnings of, 195–96

behaving like a business, 196, 201, 223

and Content Liberation Front, 193–94, 201

downloaded material returned to, 7, 231, 236

and MIT, 197–201, 215–16, 256

and Swartz's court case, 222, 231–32, 235

Swartz's downloads from database of, 1, 3, 196–202, 207, 213, 215, 222, 228, 235, 256

terms of service, 197, 198, 202, 207, 217–18

Kaestle, Carl F., 69

Kafka, Franz, *The Trial*, 221–22, 244–45

Kahle, Brewster, 135–36, 162–63, 242

Kaminstein, Abraham, 84

Kan, Gene, 133–34

Karnofsky, Holden, 234

Karp, Irwin, 84

Kastenmeier, Robert W., 90
Keating, Kenneth B., 88
Keker & Van Nest, 6, 254
Kendall, Joshua, 21
Kennedy, John F., 79, 80
King James Bible, 105
Kirk, Mark, 242
Kling, Rob, 13, 152
Knappenberger, Brian, 14
knowledge:
    copyright as tax on, 46–47, 49, 61
    digital dissemination of, 11
    organizing all of (universal brain),
        90–91
    and power, 172
    public access to, 4, 5, 13, 39, 80,
        81, 99–100
    retarding the spread of, 121
    sharing of, 4, 101, 103, 172
    social mobility via, 48
    unauthorized access to, 13–14
Koman, Richard, 135–36
Kottke, Jason, 131

Laham, Tim, 215
LaMacchia, David, 119–20
LaMacchia loophole, 119
Lamb, Roberta, 13, 152
language:
    and education, 48–49
    and national identity, 22–23, 28,
        40
Larsen, Brian, 198–99
Lassila, Ora, 128
Leahy, Patrick, 225, 226, 230, 231,
    242
Leslie, Frank, 52, 55
Leslie, Stuart W., *The Cold War and
    American Science*, 209

Lessig, Lawrence, 3, 121–23, 143,
    252
    changing the world, 185
    and copyright, 129, 134
    and Creative Commons, 129–31
    and *Eldred v. Ashcroft,* 122–23,
        129, 134, 136–41
    and Emerging Technologies
        Conference, 131–32, 136
    Free Culture, 140
    The Future of Ideas, 132
    at Harvard, 205
    political corruption as focus of,
        172, 184
    at Swartz's memorial, 262
    and W3C, 129
Leverenz, David, 38
Levy, Steven, *Hackers,* 102–3, 138
libraries:
    academic, 175
    attraction of, 99–100
    automated, 82, 90, 91
    budgets of, 175–76, 215
    in developing countries, 173–74,
        175–76, 177–78
    digitized, 82–83, 101, 109, 113,
        135
    Electronic Information for
        Libraries (EIFL), 173–74, 179,
        190
    federal depository, 187
    of the future, 13, 81–83, 101, 269
    as gift economy, 89
    Google Print for Libraries, 163
    infinite, 3, 13, 163, 195
    Memex (linked-information
        retrieval system), 82–83, 108
    as national symbol, 80
    and OCLC, 179–80

"one big library," 90–91, 127–28, 162–63
public, 67–69, 70, 80, 81, 100, 162
volunteer librarians in, 120
Library of Alexandria, 135, 139, 162
Library of Congress, 69, 70–71, 77–78, 139
Library/USA, 80–81, 82
Licklider, J. C. R. "Lick," *Libraries of the Future*, 101
literature:
American writers, 52
audiences for, 51–52, 65
British, 56, 67
e-books, 99, 107, 117
and marketability, 65
as property, 65, 70
public-domain, 99
quality of, 50, 51, 65
unauthorized reprints, 42–43, 53, 56
Lodge, Henry Cabot, 62, 63
Lovell, John W., 56, 57, 58
Lowell, James Russell, 60
Ludlum, Robert, 177
Lyon, Matthew, 101

Maclaurin, Richard, 207–8
Malamud, Carl, 181–93
Exploring the Internet: A Technical Travelogue, 182–83
open-data advocacy of, 183–85
and PACER, 185–93
and Swartz, 188–93, 222, 223
Marcus Aurelius, 99
marketing, push vs. pull approach to, 133–34, 268
Marryat, Capt. Frederick, 41–44, 47, 54, 269

Diary in America, 44, 45–46
Mr Midshipman Easy, 41, 42
Snarleyyow, 43
Marx, Karl, 69
Mary, queen of England, 18
Massachusetts Institute of Technology (MIT):
Artificial Intelligence (AI) Lab, 102–4, 112, 125, 135, 212, 213
corporate-funded research in, 210–13, 265
government contracts in, 208–9
hacker attack on, 262–63
hacker ethic at, 206, 212
International Puzzle Mystery Hunt Competition, 204, 227, 270
and JSTOR, 197–201, 215–16, 256
and library of the future, 81–82
Media Lab, 203–4, 212–13, 219, 232, 250, 270
and Project Intrex, 82, 83, 90, 91
reputation for openness in, 206–7, 213, 256
Swartz barred from, 222, 227
and Swartz's death, 262, 264–65
Swartz's downloads from, 1, 3, 197–202, 207, 213, 215, 235
and Swartz's legal woes, 232, 249–50, 254, 262, 264–65
as technical institute, 207–8
Mathews, Cornelius, 50–53
Mayersohn, Jeffrey, 252, 257
McCollum, Bill, 119
McGill, Meredith L., 31
McGrane, Reginald, 44
McKim, Charles Follen, 67

McLean, Sam, 258
McLeod, Kembrew, 48
McSherry, Corynne, *Who Owns Academic Work?*, 88–89
Mead Data Central, 184, 185
Megaupload, 226
Melville, Herman, 52
Memex, 82–83, 108
Menendez, Bob, 242
Menken, Alan, 110–11
Merkley, Jeff, 242
Merton, Robert K., "The Normative Structure of Science," 209–10, 212
Milic, Louis T., 96
Miller, Perry, *The Raven and the Whale*, 51
mimeograph, 69, 88
Mosaic, 108
Moses, Robert, 203
Motion Picture Association of America (MPAA), 111, 239, 243
motion pictures, 69
movable type, invention of, 18
MoveOn, 258
MS-DOS, 106
Mumford, L. Quincy, 84
Murdock, Georgia Ann, 85
Murphy, Joseph, 215
Murphy, Lawrence Parke, 55
music industry:
    and copyright law, 72–73, 74–76, 84, 85, 87, 110–11
    and iTunes, 153, 178
    major record labels, 154
    and Napster, 133
    new technologies in, 69, 71, 87, 133, 152–53
    pay what you can, 154
    peer-to-peer file sharing, 133, 152–54
    piracy in, 73, 87, 134, 152–54
    public-performance clause, 87
    royalties in, 77, 87
Music Publishers' Association, 72

Napster, 133
National Endowment for the Humanities, 117
National Science Foundation (NSF), 82
Nature Conservancy, The, 129
Neuefeind, Bettina, 252
newspapers:
    cost of newsprint, 55
    "Golden Age of Newspaper Hoaxes," 48
    illustrated, 52
    popularity of, 69
    postage rates for, 31, 34
    primary source of cultural information, 31–32, 48, 49, 51
    propaganda in, 60, 62
    syndicated material in, 60
    uncopyrighted, 31
newsstands, 55
Newton, Isaac, 210
New York World's Fair (1964), 79–81
No Electronic Theft Act (NET Act), 119, 132–33, 237
nonprofit organizations, inefficiencies of, 259
North American Review, 35, 70, 71
Norton, Ada, 258
Norton, Quinn, 174
    meeting of Swartz and, 131–32
    and Open Access Manifesto, 180–81

subpoenaed by prosecutors, 1–2, 227–29
and Swartz's arrest, 217, 222
Swartz's relationship with, 2–3, 9–10, 161, 229, 234
Not a Bug, 150–51, 156, 158–59, 164

O'Brien, Danny, 131
O'Connell, John J., 76
Ohanian, Alexis, 149–50, 158–59
Omidyar Network, 185
Online Book Initiative, 112
Online Computer Library Center (OCLC), 179–80
open access, 174–81, 189–90, 230
Open Content Alliance, 162–64
open-information movement, 98, 103
Open Library, 163, 173, 179, 223, 228, 267
O'Reilly Media, 129, 131
Ortiz, Carmen M., 4–5, 264
Oyama, Katherine, 239

Pallante, Maria, 239
Panic of 1837, 44
papermaking, wood-pulp, 55
Pasco, Samuel, 63, 64
Patel, Amit, 165
Patry, William, 26, 31
    Moral Panics and the Copyright Wars, 133
penny press, 48
Perault, Jay, 216, 217
Peters, Elliot, 6, 254, 256–57, 261, 263
Peters, Marybeth, 120
Peterson, John R., 86
Philadelphia, publishers in, 47

photocopier, 87–88
piano rolls, 71, 75
Pickering, John, 35
Pickett, Michael, 215, 216, 217
Pierce, Albert, 217
Pinterest, 241
pioneering spirit, 112
PIPA (2011), 230–31, 237–44, 247, 248
piracy, 86–87
    and cheap books, 56, 58, 61
    and hackers, 104
    and legislation, 50, 61, 120, 133, 225–27
    and literacy, 39
    in music industry, 73, 87, 134, 152–54
    online infringement, 133
    in schools, 88
    and sharing, 4, 133, 137, 152–54, 179
    Stop Online Piracy Act, 238–44
    unauthorized access as, 13–14
    of works by non-US authors, 39, 42–43, 53, 268
Poe, Edgar Allan, 52
Poe, Ted, 239, 242
Posner, Richard, 121
postal system, 31, 34, 56
Post Office Act (1792), 31
Pound, George W., 74, 76
Preventing Real Online Threats, *see* PIPA
printing press:
    development of, 18, 98–99
    steam-driven, 48
printing technology, 18, 48, 69
Progressive Change Campaign Committee (PCCC), 202–3, 225

Project Gutenberg, 99, 105–7, 108–9, 112–15, 117, 118, 237, 242
Project Intrex, 82, 83, 90, 91
property laws:
    and copyright, 5, 11, 26, 39, 65, 223, 268
    and intellectual property, 13–14, 88
Protestant Reformation, 99
Public Access to Court Electronic Records (PACER), 185–93, 202, 223, 267
public domain:
    archiving all books in, 135, 162, 163, 173
    challenge to legitimacy of, 86, 140
    classic titles in, 118
    and Congress, 111–12, 132–34
    and copyright laws, 3–4, 86, 115
    and copyright term, 4, 110–11, 134–35
    and *Eldred v. Ashcroft*, 122–23
    and free culture movement, 3–4
    as penalty vs. opportunity, 115
    and public access, 185–87
    purpose of, 112
    and social value, 4, 135
Public Knowledge, 230
public.resource.org, 185
Publishers' Weekly, 53, 56, 58, 59
publishing:
    of academic research, 175–77, 178
    as "best-seller system," 65
    commercial viability of, 13, 25–26, 39, 121, 175
    "courtesy of the trade," 54–56, 65
    electronic, 120
    and invention of movable type, 18

    of non-US books, 39, 41, 46–47
    percentage of authors' royalties to, 41
    protectionist laws, 120
    serials pricing crisis in, 175
    of unauthorized editions, 42–43, 53, 56
    white-shoe East Coast, 54–55
Putnam, George Haven, 53–55, 56, 57
    and free public libraries, 70
    and international copyright, 53, 59, 60, 64
    Memories of a Publisher, 54
Putnam, George Palmer, 45
Putnam, Herbert:
    as Boston head librarian, 67, 70
    and copyright law revisions, 72–73, 75–76
    death of, 78
    as librarian of Congress, 70–71, 77–78
    on public libraries, 80, 100
Putnam, John, 71
Putnam's, 52

Quine, Willard Van Orman, 217

Radway, Janice A., 69
Ramsay, David, 25, 34
Rand, Ayn, *Atlas Shrugged,* 107
RAND Corporation, 82
Raw Nerve, 251–53, 255
    ask others for help, 252–53, 257
    believe you can change, 251–52
    confront reality, 254
    lean into the pain, 252, 257
    on systemic failure, 265
    take a step back, 252

reading:
    cheap books, 52, 55–56, 58, 59, 61
    dime novels, 52, 55
    e-books, 99, 107, 117
    escapism in, 52
    literacy rates, 25, 26–27, 39, 44, 48
    penny press, 48
    value of, 48–49
recorded sound, 69, 71, 74, 77
Recording Industry Association of
    America (RIAA), 134, 152–53
Reddit, 156–61, 164, 223
    development of, 149–51
    and Internet Censorship Day, 240,
        241
    sale of, 2, 156, 158, 170
    Swartz's departure from, 159–61,
        171, 248
Reed Elsevier, 175, 178–79, 180, 239
Reformation, 99
Rehnquist, William, 138
Rein, Lisa, 123, 130, 139, 141, 269
Remember Aaron Swartz, 261
Rensselaer, Stephen van, 35
resource.org, 187
Reville, Nicholas, 152
robotic harvesting, 198–99
Romuald (monk), 169
Roosevelt, Franklin D., 78, 82, 208
Roosevelt, Theodore, 70, 75
Rousseau, Jean-Jacques, 151
Rules, The, breakers of, 14
Rush, Benjamin, 33
Russell, Bertrand, 254
Ryshke, Robert, 127

Sadler, Bess, 181
Santana, Carlos, 111
Scalia, Antonin, 121

Scheiber, Noam, 201
Schoen, Seth, 8, 138–39, 144, 148
Schonfeld, Roger, *JSTOR*, 195, 196
Schoolyard Subversion (blog), 126
Schulman, John, 85
Schultze, Stephen, 189
Schwartz, John, 191
Schweber, S. S., 208–9
Scialabba, George, 5
scientific ethos, 210
scientific research:
    community ownership of, 210
    corporate funding of, 209–13
    government funding of, 82–83,
        101, 174
    military funding of, 209
    serials pricing crisis in, 175
Scoble, Robert, 131
Scott, Sir Walter, 42
Scribner, Charles, 61, 63
Scrutton, Sir Thomas, 84
Scudder, Horace, 23–24
Securities and Exchange Commission
    (SEC), 183–85
Segal, David, 225, 231, 236, 238
Semantic Web, 267
Seuss Enterprises, 134
Shepard, Alan, 78
Shove, Raymond, 58
Silicon Valley, 152, 157, 161, 165,
    167, 267
Simon, Taryn, 244
Simonds, William E., 61
Singel, Ryan, 229
Smith, Lamar, 238
Snyder, K. Alan, 34
Sonny Bono Copyright Term
    Extension Act (CTEA, 1998),
    118–20, 122, 139, 172, 269

Soros, George, 201
Sousa, John Philip, 74–75, 77
stagecoaches, postal mail delivered
    by, 31
Stallman, Richard, 102–5, 107, 190,
    268
Stamos, Alex, 256
Stanford University, 143–45
Stationers' Company, England, 18–19
Statute of Anne, 19–20, 30–31
Stearns, Richard, 120
Stedman, Edmund Clarence, 64
STEM journals, 175
stereotype printing, 48
Stern, Madeleine B., 52, 55
Stevenson, Louise, 48–49
Stinebrickner-Kauffman, Taren:
    and Swartz's death, 261, 262
    and Swartz's legal woes, 3, 5, 234,
        249, 250–51, 257
    Swartz's relationship with, 1,
        6, 9–10, 214, 233–34, 241,
        243–44, 252, 258, 260
Stop Online Piracy Act (SOPA,
    2011), 238–44, 248
Stowe, Harriet Beecher, 52
Summer Founders Program, 147,
    148, 149, 204
Summers, John, 5
Sundance Film Festival, 14
Supreme Court, US:
    computer system of, 121
    Eldred v. Ashcroft, 122–23, 124,
        129, 134, 136–41, 242
Swartz, Aaron, 123–32
    on A/B testing the revolution, 7,
        215
    arrest of, 1, 5, 7, 222, 229; *see
        also* Swartz's legal case

as "big hacker," 223, 237
blogs of, 3, 126, 137, 140, 144, 155,
    160, 162, 166, 204, 206, 251
Bubble City, 164–65
and copyright reform movement,
    124
and Creative Commons, 130
in Cupramontana, 170, 173–74,
    177–78
death of, 10, 11, 14, 261–66
and *Eldred v. Ashcroft,* 137–38,
    140
FBI file on, 191–92, 223
fleeing the system, 8, 145, 151,
    158–59, 161, 171, 173, 193,
    248, 267
and free culture movement, 3–4,
    141, 152–55, 167, 223
and Harvard, 3, 205, 207, 223,
    224, 229
health issues of, 9, 150, 165–66,
    222
immaturity of, 8–9
and Infogami, 147, 148–51, 158
interests of, 6–7, 8–9, 204, 221
"Internet and Mass
    Collaboration, The," 166–67
lawyers for, 6, 254–55
legacy of, 14–15, 268, 269–70
and Library of Congress, 139
and Malamud, 187–93, 222, 223
manifesto of, 6–7, 178–81,
    189–90, 201, 228–30, 247
mass downloading of documents
    by, 1, 3, 188–94, 197–202,
    207, 213, 215, 222, 228, 235
media stories about, 125
and MIT, 1, 3, 201, 204, 207, 213,
    222, 227, 232, 249–50, 262

and money, 170–71
on morality and ethics, 205–6
and Open Library, 163, 173, 179,
    223, 228
and PCCC, 202–3, 225
as private person/isolation of,
    2–3, 5, 124, 127, 143, 154–55,
    158–60, 166, 169, 205, 224,
    227, 228, 248–49, 251
and public domain, 123
as public speaker, 213–14, 224,
    243, 257
and Reddit, see Reddit
The Rules broken by, 14
"saving the world" on bucket list
    of, 7, 8, 15, 125, 151–52, 181,
    205–6, 247–48, 266, 267, 268
self-help program of, 251–53
and theinfo.org, 172–73
and US Congress, 224–25,
    239–40
Swartz, Robert:
    and Aaron's death, 261, 262, 264
    and Aaron's early years, 124, 127
    and Aaron's legal woes, 232, 250,
        254
    and MIT Media Lab, 203–4, 212,
        219, 232, 250
    and technology, 124, 212
Swartz, Susan, 128–29, 160, 192
Swartz's legal case:
    as "the bad thing," 3, 7–8, 234
    change in defense strategy,
        256–57
    evidence-suppression hearing,
        259–60
    facts of, 11
    felony charges in, 235, 253
    grand jury, 232–33

indictment, 1, 5, 8, 10, 11, 233,
    234, 235–37, 241, 253–54
investigation and capture,
    215–17, 223, 228
JSTOR's waning interest in,
    231–32
manifesto as evidence in, 228–30
motion to suppress, 6
motives sought in, 223, 229
Norton subpoenaed in, 1–2,
    227–29
ongoing, 248, 249–51
online petitions against, 236–37
original charges in, 218, 222
plea deals offered, 227, 250
possible prison sentence, 1, 2, 5,
    7–8, 11, 222, 232, 235–36,
    253, 260
potential harm assessed, 218, 219,
    222, 235
prosecutor's zeal in, 7–8, 11, 218,
    222–24, 235–37, 253–54,
    259–60, 263, 264
search and seizure in, 6, 223–24,
    256–57
Symbolics, 103
systems, flawed, 265–67

T. & J. W. Johnson, 49
Tammany Hall, New York, 57
tech bubble, 146, 156
technology:
    Bayesian statistics in, 258–59
    burgeoning, 69, 71, 84, 87–88
    communication, 12, 13, 18, 87–88
    computing, see computers
    and digital culture, 122
    and digital utopia, 91, 266–67
    of electronic publishing, 120

technology (*cont.*)
  and intellectual property, 90–91
  and irrational exuberance, 146
  in library of the future, 81–83
  as magic, 152
  moving inexorably forward, 134
  overreaching police action
    against, 233
  power of metadata, 128, 130
  as private property, 210
  resisting change caused by, 120
  saving humanity via, 101
  thinking machines, 102
  unknown, future, 85
  and World War II, 208
telephone, invention of, 69
Templeton, Brad, 261
theinfo.org, 172–73
theme parks, 134
ThoughtWorks, 9, 248, 257, 258
"thumb drive corps," 187, 191, 193
Toyota Motor Corporation, "lean
    production" of, 7, 257, 265
Trumbull, John, *McFingal,* 26
trust-busting, 75
Tucher, Andie, 34
Tufte, Edward, 263–64
"tuft-hunter," use of term, 28
Tumblr, 240
Twain, Mark, 60, 62, 73
Tweed, William "Boss," 57
Twitter, 237

Ulrich, Lars, 133
United States:
  Articles of Confederation, 26
  copyright laws in, 26–27
  economy of, 44–45, 51, 55, 56
  freedom to choose in, 80, 269

industrialization, 57
  literacy in, 25, 26–27, 39, 44, 48
  migration to cities in, 57
  national identity of, 28, 32
  new social class in, 69–70
  opportunity in, 58, 80
  poverty in, 59
  railroads, 55, 56
  rustic nation of, 44–45
  values of, 85
UNIVAC computer, 81, 90
Universal Studios Orlando, 134
University of Illinois at Urbana-
    Champaign, 94, 95–96, 112–15
Unix, 104
US Chamber of Commerce, 239
utilitarianism, 214

Valenti, Jack, 111, 132
Van Buren, Martin, 44
Van Dyke, Henry, *The National Sin
    of Literary Piracy,* 61
venture capital, 146
Viaweb, 146
Victor, O. J., 59
Victory Kit, 258–59
Vixie, Paul, 185
voting rights, 57, 58
Voyage to America (film), 79

Walker, Scott, 233
Wallace, David Foster, 6, 159, 244
Walton, Sam, 7, 249
WarGames (movie), 218
Washburn, Jennifer, *University, Inc.,*
    212
Washington, George, 29–30
watchdog.net, 173, 188, 191, 193, 257
web.resource.org, 187–88

Webster, Abraham, 36
Webster, Daniel, 36
Webster, Noah, 20–25, 268
  ambition of, 24, 27–28
  American Dictionary of the
    English Language, 34, 35–37
  as author, 23, 24–25
  autobiography of, 29–30
  "blue-backed speller" by, 23, 25,
    27, 30, 33, 34, 36
  and copyright, 23–24
  and copyright law, 27–31, 36–40
  death of, 40
  early years of, 21
  as Federalist, 32–33
  financial problems of, 33, 36
  lobbying by, 27–30, 37
  public image of, 38
  as public speaker, 28
  as teacher, 22
Webster, William, 37
Weinberg, Martin, 254, 256
Weinberger, David, 158
Wells, H. G., 99
Westlaw database, 173
White Friars, 174
Whole Earth Catalog, 12
Wikimedia Foundation, 150
Wikipedia, 124, 173, 241
Wikler, Ben, 10, 233, 248–49, 260
  and Avaaz, 203, 241, 248
  and *Flaming Sword of Justice,*
    241, 243
  and Swartz's death, 262
  Swartz's friendship with, 203,
    214, 262
  and Swartz's legal woes, 202, 217
Wilcox-O'Hearn, Zooko, 125, 129,
  145

Wilhelm II, Kaiser, 71
Williams, Julie Kay Hedgepeth, 27
Wilson, Christopher P., *The Labor of*
  Words, 64–65, 70
Wilson, Holmes, 152, 155, 240, 241,
  243, 263
Windows operating system, 106
Winer, Dave, 8, 131
Winn, Joss, 207
Wolcott, Oliver Jr., 25, 32, 33
Woodhull, Nathan, 10, 259
WordPress, 241
work as identity, 146
WorldCat, 179
World War I, 77
World War II, 78, 82, 208
World Wide Web:
  anniversary of, 237–38
  archiving all of, 135–36, 173
  commercial potential of, 112
  as infinite library, 127–28
  and Internet, 98, 108–10
  introduction of, 98, 108
  linking capacity of, 108, 238
  malignant forces vs., 238
  open, collaborative, 178, 237
  popularization of, 112
World Wide Web Consortium
  (W3C), 127–29
Wyden, Ron, 226, 231

Xerox photocopy machine, 87–88
Xerox Sigma V mainframe, 95–97, 113

Yahoo, 185
Y Combinator, 147
Young America, 50–53

Zanger, Jules, 44

ABOUT THE AUTHOR

Justin Peters is a correspondent for *Slate* and a contributing editor at the *Columbia Journalism Review*. His essay on Peter Fleming's book *Brazilian Adventure* was anthologized in *Second Read: Writers Look Back at Classic Works of Reportage*. He divides his time between Boston and New York.